T0203022

Communications
in Computer and Information Science 2081

Editorial Board Members

Joaquim Filipe , *Polytechnic Institute of Setúbal, Setúbal, Portugal*
Ashish Ghosh , *Indian Statistical Institute, Kolkata, India*
Raquel Oliveira Prates , *Federal University of Minas Gerais (UFMG),*
Belo Horizonte, Brazil
Lizhu Zhou, *Tsinghua University, Beijing, China*

Rationale
The CCIS series is devoted to the publication of proceedings of computer science conferences. Its aim is to efficiently disseminate original research results in informatics in printed and electronic form. While the focus is on publication of peer-reviewed full papers presenting mature work, inclusion of reviewed short papers reporting on work in progress is welcome, too. Besides globally relevant meetings with internationally representative program committees guaranteeing a strict peer-reviewing and paper selection process, conferences run by societies or of high regional or national relevance are also considered for publication.

Topics
The topical scope of CCIS spans the entire spectrum of informatics ranging from foundational topics in the theory of computing to information and communications science and technology and a broad variety of interdisciplinary application fields.

Information for Volume Editors and Authors
Publication in CCIS is free of charge. No royalties are paid, however, we offer registered conference participants temporary free access to the online version of the conference proceedings on SpringerLink (http://link.springer.com) by means of an http referrer from the conference website and/or a number of complimentary printed copies, as specified in the official acceptance email of the event.

CCIS proceedings can be published in time for distribution at conferences or as postproceedings, and delivered in the form of printed books and/or electronically as USBs and/or e-content licenses for accessing proceedings at SpringerLink. Furthermore, CCIS proceedings are included in the CCIS electronic book series hosted in the SpringerLink digital library at http://link.springer.com/bookseries/7899. Conferences publishing in CCIS are allowed to use Online Conference Service (OCS) for managing the whole proceedings lifecycle (from submission and reviewing to preparing for publication) free of charge.

Publication process
The language of publication is exclusively English. Authors publishing in CCIS have to sign the Springer CCIS copyright transfer form, however, they are free to use their material published in CCIS for substantially changed, more elaborate subsequent publications elsewhere. For the preparation of the camera-ready papers/files, authors have to strictly adhere to the Springer CCIS Authors' Instructions and are strongly encouraged to use the CCIS LaTeX style files or templates.

Abstracting/Indexing
CCIS is abstracted/indexed in DBLP, Google Scholar, EI-Compendex, Mathematical Reviews, SCImago, Scopus. CCIS volumes are also submitted for the inclusion in ISI Proceedings.

How to start
To start the evaluation of your proposal for inclusion in the CCIS series, please send an e-mail to ccis@springer.com.

Sree Lakshmi Gundebommu ·
Lakshminarayana Sadasivuni ·
Lakshmi Swarupa Malladi

Editors

Renewable Energy, Green Computing, and Sustainable Development

First International Conference, REGS 2023
Hyderabad, India, December 22–23, 2023
Proceedings

Editors
Sree Lakshmi Gundebommu (iD)
CVR College of Engineering
Hyderabad, Telangana, India

Lakshminarayana Sadasivuni (iD)
Andhra University
Visakhapatnam, Andhra Pradesh, India

Lakshmi Swarupa Malladi (iD)
CVR College of Engineering
Hyderabad, India

ISSN 1865-0929   ISSN 1865-0937 (electronic)
Communications in Computer and Information Science
ISBN 978-3-031-58606-4   ISBN 978-3-031-58607-1 (eBook)
https://doi.org/10.1007/978-3-031-58607-1

© The Editor(s) (if applicable) and The Author(s), under exclusive license
to Springer Nature Switzerland AG 2024

This work is subject to copyright. All rights are solely and exclusively licensed by the Publisher, whether the whole or part of the material is concerned, specifically the rights of translation, reprinting, reuse of illustrations, recitation, broadcasting, reproduction on microfilms or in any other physical way, and transmission or information storage and retrieval, electronic adaptation, computer software, or by similar or dissimilar methodology now known or hereafter developed.
The use of general descriptive names, registered names, trademarks, service marks, etc. in this publication does not imply, even in the absence of a specific statement, that such names are exempt from the relevant protective laws and regulations and therefore free for general use.
The publisher, the authors and the editors are safe to assume that the advice and information in this book are believed to be true and accurate at the date of publication. Neither the publisher nor the authors or the editors give a warranty, expressed or implied, with respect to the material contained herein or for any errors or omissions that may have been made. The publisher remains neutral with regard to jurisdictional claims in published maps and institutional affiliations.

This Springer imprint is published by the registered company Springer Nature Switzerland AG
The registered company address is: Gewerbestrasse 11, 6330 Cham, Switzerland

Paper in this product is recyclable.

Preface

We are very delighted to introduce the proceedings of the International Conference on Renewable Energy, Green Computing & Sustainable Development (REGS 2023), which was held at the CVR College of Engineering, Ibrahimpatnam, Hyderabad, India during 22nd – 23rd December 2023. The theme of the conference was "Renewable Energy, Green Computing and Sustainable Development". We were grateful to all distinguished participants for their contribution to the conference. Participants from 8 countries and 2 continents (Europe and Asia) presented and shared their work on a variety of topics. A total of 133 papers were received through the CMT platform from different geographical parts of the world: Dubai, Ecuador, Mongolia, Oman, Czech Republic, Ukraine, Republic of Korea, Iraq, and many states of India. After a double-blind peer review with three reviewers per paper, 15 papers were selected for this Springer CCIS publication. The full manuscripts were selected based on originality, significance, relevance, and technical soundness. The two-day conference had five keynote lectures by eminent personalities. The oral sessions were conducted offline on the first day and online through the Zoom platform on the second day with three different tracks organized in six sessions.

It is good to have an opportunity to mention our invited speakers Satish Kumar Peddapelli, IIIT Basara & Osmania University, India, Lakshminarayana Sadasivuni, Andhra University, India, Milan Belik, University of West Bohemia, Czech Republic, Olena Rubanenko, University of West Bohemia, Czech Republic, and Sarat Kumar Sahoo, Parala Maharaja Engineering College, India, all of whom delivered keynote addresses and invited talks. Our special thanks go to our speakers and all the authors for sharing their latest research findings.

We would like to acknowledge and thank the conference committee members and international reviewers who spent a lot of time on the arduous task of selecting the qualified papers from all the submissions. Their excellent professional advice enabled us to come out with this volume of proceedings of the conference.

All in all, we would like to express our sincere thanks to all participants for their contributions to the conference and also appreciate the hard work of the committee in organizing the conference. We strongly believe that this conference was an engrossing and captivating discussion for every participant. We thank the CVR College of Engineering management, who encouraged us morally and financially.

January 2024

Sree Lakshmi Gundebommu
Lakshminarayana Sadasivuni
Lakshmi Swarupa Malladi

Organization

General Chair

G. Sree Lakshmi   CVR College of Engineering, India

Program Committee Chairs

Jih-Sheng (Jason) Lai	Virginia Tech., USA
Ned Mohan	University of Minnesota, USA
K. Lal Kishore	CVR College of Engineering, India
S. Lakshminarayana	Andhra University, India
Tushara Sadasivuni	Georgia State University, USA
Sudarsan Sadasivuni	Intel Corporation, USA
Satyam Paul	University of Nottingham, UK
Olena Rubanenko	University of West Bohemia, Czech Republic
S. Venkateshwarlu	CVR College of Engineering, India
K. Shashidhar Reddy	CVR College of Engineering, India
M. Lakshmi Swarupa	CVR College of Engineering, India
Ch. Lokeshwar Reddy	CVR College of Engineering, India
P. Uma Maheshwera Reddy	CVR College of Engineering, India

Steering Committee

Hamdan Bin Daniyal	University Malaysia Pahang, Malaysia
Zi-Qiang Zhu	University of Sheffield, UK
Md. Kamrul Hasan	University of Engineering and Tech., Bangladesh
Saad F. Al-Gahtani	King Khalid University, Saudi Arabia
Mofeed Turky Rashid	University of Basrah, Iraq
Amitava Roy	University of Portsmouth, UK
Mansour S. Farhan	Wasit University, Iraq
Md. Shafiqul Islam	BUET, Bangladesh
Vidya Sagar Yellapu	University of Huddersfield, UK
Surender Reddy Salkuti	Woosong University, South Korea
J. B. Ekanayake	University of Wollongong, Australia
George Ghinea	Brunel University London, UK
Selwyn Piramuthu	University of Florida, USA

Sherali Zeadally	University of Kentucky, USA
Tarun Kumar Lohani	Arba Minch University, Ethiopia
Akhtar Kalam	Victoria University, Australia
Hasan Fahad Khazaal	University of Wasit, Iraq
Oleksandr Rubanenko	Vinnitsia National Technical University, Ukraine
Iryna Hunko	Vinnitsia National Technical University, Ukraine
Peter Lezhnyuk	Vinnitsia National Technical University, Ukraine
Pierpaolo Greco	University of Modena, Italy
Chau Yuen	Nanyang Technological University, Singapore
Richi Nayak	Queensland University of Technology, Australia
Xiao-Zhi Gao	Aalto University, Finland
Sigurd Meldal	San José State University, USA
Syed. M. Buhari	King Abdulaziz University, Saudi Arabia
Xavier Fernando	Ryerson University, Canada
Zairi Ismael Rizman	University of Tecknologi MARA, Malaysia
Dibakar Das	University of Hyderabad, India
Alivelu Manga Parimi	BITS Pilani, India
Parikshit Sahatiya	BITS Pilani, India
D. M. Vinod Kumar	NIT Warangal, India
T. Vineshwaran	VIT, Chennai, India
Jacob Raglend	VIT, Vellore, India
D. Susitra	Sathyabama Institute of Science & Technology, India
Muralidharnayak Bhukya	Central University of Haryana, India
Tejavathu Ramesh	NIT Andra Pradesh, India
K. A. Chinmaya	IIT(BHU) Varanasi, India
Gaddam Tulasi Ram Das	JNTUH, India
Indra Nayankar	IIT Delhi, India
Sanket Goel	BITS Pilani, India
Tanuja Sheoreyn	PDPM IIITDM, Jabalpur, India
D. Godwin Immanuel	Sathyabama University, India
Trapti Jain	IIT Indore, India
S. Jayaprakash	Sathyabama University, India
Olive Ray	IIT Bhubaneswar, India
Yogesh Vijay Hote	IIT Roorkee, India
Srinivasan Thiruvenkadam	VIT Vellore, India
S. Meikandasivam	VIT Vellore, India
Amitesh Kumar	NIT Patna, India
Mummadi Veerachary	IIT Delhi, India
Kodeeswara Kumaran G.	MSRIT, India
Debapriya Das	IIT Kharagpur, India
Ankit Kumar Singh	NSUT, India

Jitendra Kumar   NIT Jamshedpur, India
Giribabu Dyanamina   MANIT Bhopal, India
Arvind Kumar Prajapati   NIT Jamshedpur, India
S. D. Sundarsingh Jebaseelan   SICT, India
B. V. Sumangala   Ramaiah Institute of Technology, India
N. Vishwanathan   NIT Warangal, India
Avik Battacharya   IIT Roorkee, India
Chandrasekhar Perumalla   IIT Bhubaneswar, India
Om Hari Gupta   NIT Jamshedpur, India
B. Chitti Babu   IIITDM Kancheepuram, India

Program Committee

A. S. S. Murugan   CVR College of Engineering, India
R. Naveena Bhargavi   CVR College of Engineering, India
G. Janardhan   CVR College of Engineering, India
Ch. Shravani   CVR College of Engineering, India
G. Divya   CVR College of Engineering, India

Additional Reviewers

A. V. Ravi Teja   B. Kumar
Abdul Saleem Shaik   Bindu Madhavi Garine
Abhishek Potnis   Gobinda Acharya
Adithyakashyap   Himani Sharma
Ajay Bansal   R. S. Ch. Murthy Chodisetty
Ankit Kumar Singh   Sengathir J.
Ankita Awasthi   Sushma Gupta
Athimamula Srujana   Ajay Kumar
B. Dhanalaxmi   B. Nagi Reddy
B. Naga Malleswara Rao   D. Hari Krishna
Besta Hariprasad   Kumar K.
Canavoy Narahari Sujatha   Deepa K.
Ch. Sudhakar   K. Sedhu Raman
Chakradar Adupa   K. Ranjith Kumar
Chandrasekhar Perumalla   Katuri Rayudu
Chitti Babu   Ranjan Kumar Behera
Chitumodhu Bargava   Vigneswaran
Deepika Rani Nagumalli   G. V. Nagesh Kumar
Dhinesh Balasubramanian   Giri Babu
Abhishek Tripathi   GopiKrishna Pasam

Prabhat Singhnit
Prakash K. Ray
Praveen Chakkravarthy
Praveen J.
Praveen Madiraju
Prince Winston
Priyanka Chawla
Ramamohan Rao
Rahul Jammy
Rajarshi Mahapatra
Rajasekhar Varma
Rajeswaran Nagalingam
Rajgopal V.
Rama Rao Pokanati
Ravi Kumar Avvari
Ravindra Janga
Saad Al-Gahtani
Santhosh Kumar Yat
Sarat Sahoo
Saravanan S.
Satyanarayana Nimmala
Seethalekshmi K.
Selvendran
Shankaraiah Ningaiah
Harish Pulluri
Harshit Bhavsar
Hemachandu Pattur
Hemant Saxena
Hima Bindu
Iryna Hunko
Jagadeeswara Rao Annam
Jaiganesh Mahalingam
Jhansi Bharathi Madavarapu
Jitendra Kumar
K. Sudha
Kalpana Chaudhary
Karuppiah
Khwaja Muinuddin Chisti Md
Kishore Kumar
Kotapati Anuradha
Krishna Murthysri
Krishna Kumari Narahari
Kuldeep Saxenagla
Lakshmi Devi

Lakshminarayana Sadasivuni
Madan Kumar
Meenakshi Sundaram
Meera Viswavandya
Mullangi Pradeep
Munagala Kalavathi
Muniraju Naidu
Muthuvinayagam Madasamy
Narasimhan Venkatesh
Naveena Bhargavi R.
Nikhil Bajjuri
Nitin Dhote
Olena Rubanenko
Olive Ray
P. Srinivasa Rao Nayak
Pavan Kumar Jakkepalli
Penubarthi Sobha Rani
Phaneendra Babu Bobba
Phani K. S. V.
Phanindra Kumar Reddy
Vinod Kumar Yadav
Vishwanatha Siddhartha
Wyatt Kang
Y. Mastanamma
Yashwanth Pamu
Sudha Tushara Sadasivuni
Sudhakar Ambarapu
Surender Reddy Salkuti
Sydulu Maheswarapu
Thalluru Anil Kumar
Tulasi Ram Das
Udhayamoorthi Marannan
Vandana Patil
Vasudevan A. R.
Venugopal Dugyala
Vinay Kumar Awaar
Vinod Kumar
Sharanya M.
Sharath Kumar D. R. V. A.
S. N. Murti Sarma
Sreeramulu Mahesh G.
Srichandan Kondamudi
Subba Rao S. P. V.
Subhransu Padhee

Invited Talks

Wind and Solar Energy Applications – Technological Challenges and Advances

Satish Kumar Peddapelli

Osmania University, Hyderabad

Major issues associated with renewable energy systems, and developments in solar/wind energy equipment systems, energy storage and bioenergy applications, hybrid renewable energy systems, are elaborated as well as the measurement techniques that are used for these systems. Original research avenues in technological developments, insights to taxonomy of challenges in renewable energy applications have also been discussed. The invited talk includes case studies of implementation of solar and wind systems in remote areas.

Wind and Solar Energy Applications – Technological
Challenges and Advances

Methods of Finding the Effect of an Element in a of Larger Dimensions

Lakshminarayana Sadasivuni

College of Engineering, Andhra University, Visakhapatnam

The invited talk started with the method of the present work, which supported the user to rate an element in larger dimensions. Such studies enable for using ranking algorithms in vector functions too. An N-dimensional function is taken and measured the effect of an element. The present method applied to an additional dimension of the function i.e. N+1 dimension and to find the effect in such state. Examples are given in different areas in dynamically growing environment. Calculation of relevance and quantifying the properties such as term frequency, inverse document frequency are discussed. Disadvantages and limitations of Statistical methods, which fail in addressing these problems because of non-availability of quantitative relationship of a particular element to the group, are also explained, with advantages with this new method.

Efficiency of Hybrid Photovoltaic Systems Optimization to Fulfill the Energy Requirements of Emergency Shelters for Refugees of the Ukrainian War

Milan Belik

Department of Electrical Power Engineering, University of West Bohemia, Pilsen, Czech Republic

Hybrid photovoltaic systems which allow decreasing energy consumption for specific objects and customers at present time are cited with theory and examples. DC-coupled inverters are recognized to be more energy efficient. Some specific cases are also discussed. An example of Ukraine problem with internal migration is explained, just through constructing new communities of emergency shelters. Integrating these units into Ukrainian overloaded and damaged distribution grids are projected very precisely to limit not just power consumption but also power injection. The discussed results show that the proper choice of battery inverter technology can save interesting amount of energy produced from the installed PV system. Examples with the AC coupled system that offered not just higher flexibility and modularity, but also higher energy efficiency of the hybrid system, lower grid feed-in.

Optimisation Technologies for Hybrid PV Systems

Olena Rubanenko

Researcher of the Research and Innovation Centre for Electrical Engineering,
University of West Bohemia, Pilsen, Czech Republic

This talk discusses this phenomenon based on specific real cases that are defined by consumption profiles, battery storage system management, climate conditions, and PV system design. Simulations presented in the article demonstrate the expected annual energy flows for both technologies in a model situation. The differences between DC coupling and AC coupling solutions are explained through in-depth analyses of inverter behavior, battery behavior, charging strategies, charging losses, discharging losses, state of charge (SOC), cycle load, and the correlation between own consumption and inverter self-consumption. The case studies which were implemented are showed to choose the right battery inverter technology can lead to significant energy savings from the installed PV system. In certain cases, AC coupled systems not only offer higher flexibility and modularity but also higher energy efficiency for the hybrid system, lower grid feed-in, and better economic profitability.

IoT-Based Intelligent Smart Energy Management Systems for PV Power Generation

Sarat Kumar Sahoo

Department of Electrical Engineering, Parala Maharaja Engineering College, Government of Odisha, Sitalapalli, Berhampur

With the cleaner, more dependable, and sustainable resources, the renewable energy sector is rising quickly thus leaving many research issues. The decline in world energy use and climate change is the two most significant factors nowadays. The lecture focused on PV forecasting was essential to enhancing the efficiency of the real-time control system and preventing any undesirable effects. Solar power generation forecasting was essential for micro grid stability and security, as well as solar photovoltaic integration in a strategic approach. This lecture encouraged participants, how to use IoT, a solar photovoltaic system being monitored, and shows the proposed monitoring system is a potentially viable option for smart remote and in-person monitoring of a solar PV system.

Contents

Communications and Signal Processing

Expert Systems and Artificial Intelligence

AI Based Performance Boost in Solar PV Fuel Cell Hybrids

Pooja Soni[1]($\boxtimes$) (iD), Vikramaditya Dave[1] (iD), and Naveena Bhargavi Repalle[2] (iD)

[1] College of Technology and Engineering, MPUAT, Udaipur, India
poojaswarnakar93@gmail.com, vdaditya1000@gmail.com
[2] CVR College of Engineering, Ibrahimpatnam, Hyderabad, India
rn.bhargavi@cvr.ac.in

Abstract. Technology with renewable energy have crucial Solar energy plays a crucial role in tackling the worldwide shift towards sustainable energy sources. Photovoltaic (PV) systems and fuel cells are two prominent sources of clean energy; however, they exhibit intermittent and variable power generation patterns, hindering their widespread adoption. This paper proposes a novel approach to improve performance of Hybrids (SPV-FCH) through the integration of Artificial Intelligence (AI) techniques. The synergy aims to create more reliable, continuous power generation system with joining nature renewable energy which includes consistent contribution of fuel cells. The integration of AI algorithms offers an intelligent control mechanism that optimizes the operation of the hybrid system, thereby overcoming fluctuations in irradiance of solar, the dynamic nature of energy demand. The AI-enabled control system employs predictive analytics and machine learning algorithms to forecast solar irradiance patterns, weather conditions, and energy consumption trends. By leveraging real-time data and historical patterns, the system can dynamically adjust both the components, optimizing their performance for maximum energy output, efficiency, and overall system reliability. Furthermore, the AI system enables proactive maintenance and fault detection, enhancing the overall resilience and longevity of the hybrid system. Through continuous learning and adaptation, the AI controller refines its predictions and control strategies, ensuring optimal performance under varying environmental conditions. This paper discusses the design and implementation of the AI-enabled control system for SPV-FCH hybrids, highlighting its effectiveness in achieving improved energy yield, grid stability, and cost-effectiveness. The proposed approach not only addresses the intermittent challenges associated with solar PV but also maximizes the utilization of both technologies, contributing advancement sustainable including resilient power solutions. The findings presented in this paper contribute valuable insights into the integration of AI in renewable energy systems, paving the way for smarter and more efficient hybrid power generation technologies.

Keywords: Generation-connected energy system Renewable sources · hybrid power system · Artificial Intelligence · fuel cell · modelling · simulation

© The Author(s), under exclusive license to Springer Nature Switzerland AG 2024
S. L. Gundebommu et al. (Eds.): REGS 2023, CCIS 2081, pp. 3–16, 2024.
https://doi.org/10.1007/978-3-031-58607-1_1

1   Introduction

Multitude includes various factors are significant with complex analysis. The global Energy landscape is undergoing a significant transformation. Various factors contribute to this revolution, such as relentless growth in population, expansion surging waves of unceasing rise energy. Traditionally, the foundation of energy supply has relied on natural gas, specifically coal, oil and fossil fuels. However, solar, water, wind and geothermal energy have emerged as viable alternatives to fossil fuels. These power societies have the potential without finite resources perpetuating the depletion causing environmental degradation. Advancements in energy storage, grid management, and efficiency optimization are playing a crucial role in the development of renewable energy. As economies navigate this transition, the convergence of policy, industry, and research becomes essential in steering towards a future where clean, reliable and sustainable energy takes precedence. The narrative surrounding energy now extends beyond meeting power requirements, becoming a testament to balance, innovation, and the preservation of the planet's future. The chronicles of the past are currently being inscribed using sustainable ink, intertwined with the vision to protect the Earth's assets for the forthcoming progeny.

1. Environmental Changes: Environment will get polluted and affected due to the greenhouse gases released by combustion of fossil fuels. Release of Carbon dioxide (CO_2) causes global warming. As per IPCC (Intergovernmental Panel on Climate Change) standards, catastrophic consequences and global warning limit should be below 2 degrees Celsius above pre-industrial levels.
2. Energy: Standards, which are reliant fuels are subjected to energy security risks because of geopolitical stresses and cost fluctuations.
3. **Pollution in Environmental**: "Fossil fuel combustion not only contributes to climate change but also causes air pollution, leading to respiratory problems and other health issues. Additionally, accidents in oil drilling, transportation, and refining have resulted in environmental disasters.
4. **Depletion of Resources**: Fossil fuels are finite resources, and their extraction leads to significant environmental impacts, including habitat destruction and water pollution.

To address these challenges and build a sustainable future, there is a growing glob al consensus on the need to transition towards sustainable power generation, characterized by:

1. **Renewable Energy**: Renewable energy sources, such as solar, wind, hydro, geothermal, and biomass, offer a clean and virtually inexhaustible alternative to fossil fuels. These sources do not emit greenhouse gases during operation, reducing carbon emissions and combating climate change.
2. **Decentralization and Grid Integration**: Renewable energy technologies facilitate decentralization of energy production, allowing individuals and communities to generate their own power. Integrating renewable energy sources into the power grid enhances grid stability and resilience.
3. **Energy Efficiency**: Improving energy efficiency in various sectors, such as buildings, transportation, and industry, can significantly reduce energy consumption and associated emissions.

4. **Technological Advancements**: Advancements in energy storage, smart grids, and artificial intelligence enable more efficient management and utilization of renewable energy, addressing intermittency issues and optimizing power generation".

Hybrid power systems are like hybrid combinations for sustainable energy generation. This synergistic combination of two different renewable energy technologies overcome during the day to produce electricity through photovoltaic effect. Fuel cells are devices that utilize electrochemical reactions to generate electricity through the conversion of hydrogen or alternative fuels. Solar generation depends on daylight availability whereas consistent with minimal pollution. Moreover, excess electricity being produced with better improvement in terms of electrolysis process and same can be used in fuel cells. This is the added advantage and to have consistent reliable generation.

Based on many research methods applied and emphasized "solar PV-fuel cell hybrid power systems." A study by (Zhang et al. 2020) demonstrated that such hybrid systems can significantly improve energy efficiency and reduce carbon emissions compared to standalone PV or fuel cell systems. In addition, with energy storage, the hybrid system enhances the gid stability and renewable energy integration. These small-scale residential and major applications.

The technological synergy achieved through this integration can lead to cost savings over time, reducing reliance on grid electricity during peak demand periods and enhancing overall energy resilience (Ghenai et al., 2013; Alkhateeb et al., 2016).

Based on previous research, "the combination of solar PV and fuel cells in hybrid power systems offers a promising approach to address the challenges of intermittent renewable energy sources while leveraging the benefits of both technologies. As the world seeks to transition to a sustainable energy future, these hybrid systems hold the potential to play a vital role in achieving reliable, efficient, and environmentally friendly power generation".

Artificial Intelligence integration to "solar PV-fuel cell hybrid power systems" has some challenges and also significant opportunities. The incorporation of AI can significantly enhance the system's performance, but it also requires careful consideration of potential drawbacks (Heydari & Askarzaedh, 2016).

1.1 Objectives

- **Quality of Data and resource management**: Systems need high-quality real-time data for different components.
- **Model Complexity**: Hybrid power systems possess inherent complexity due to the involvement of various components, the presence of various predictions. The development of latest designed models capable of effectively managing offering valuable Hybrid power systems possess inherent complexity due to the involvement of multiple interacting components and the presence of unpredictable external factors. The development of AI models capable of effectively managing such complexity and offering valuable optimization insights can prove to a challenging endeavor. Prove to be a challenging endeavor.
- **Algorithm Selection:** It is a crucial task to choose proper AI algorithm for the optimized solution. Deep understanding of the capabilities and limitations of various algorithms is very much essential to choose the righ one.

- **Training and Validation**: AI model training involves large datasets and at the same time proper validation is required to generalization of the models. Suboptimal performance can only be achieved by proper training and validation.

1.2 Latest Research

- **Advanced Control**: AI can provide advanced control mechanisms for hybrid power systems, optimizing their operation and ensuring efficient utilization of resources. This can lead to cost savings and improved overall system performance.
- **Energy Market Integration**: "AI can facilitate participation in energy markets by predicting electricity prices, allowing the hybrid system to buy or sell electricity strategically, potentially leading to cost savings or revenue generation (Ghenai & Janajreh, 2016; Lambert et al., 2006; Yilmaz et al., 2015).

  - Performance of AI
  - Decision Making in Real-Time
  - Maintenance in Predictive
  - Flexibility
  - Learning Continuously

The research conducted by Zhang et al. (2020) delved into various techniques for hydrogen production, storage technologies, and the role of hydrogen in bolstering the energy storage capacity of the system. By integrating hydrogen storage into the hybrid system, the resilience of energy was enhanced, allowing for better utilization of surplus PV power. In their study, they proposed a power management method for a DC microgrid consisting of solar PV, batteries, fuel cell, and stored hydrogen. The viability of this system design was assessed using HOMER software in a remote area of North India. The findings demonstrated that the proposed structure is financially feasible, with an overall net present cost (NPC) of $83,103. Additionally, time-domain simulations were conducted to ensure the technological feasibility of the system. The results indicated that the suggested approach effectively electrifies critical loads, such as ventilators, even in the face of fluctuating solar irradiation conditions.

2 Literature Review

PV-fuel cell hybrid power systems and their performance. Here are some key findings from previous studies:

1. Techno-Economic Analysis and Optimization of Solar PV-Fuel Cell Hybrid Systems:

A study by Pang et al. (2019) conducted a techno-economic analysis of a grid-connected solar PV-fuel cell hybrid system. They found that the hybrid system reduced the overall cost of electricity generation compared to standalone PV or fuel cell systems. The optimization of the system's operation based on weather forecasts and electricity demand patterns resulted in improved energy efficiency and financial viability. (Ghenai et al., 2020) created a hybrid solar PV/fuel cell power plant to handle the electric load of a desert-area residential subdivision. The system was developed with a renewable component of 40.2% and a levelized cost of energy of 145 $/MWh in mind. The method was proven to be both economically and environmentally sustainable.

The system met the residential community's AC primary load with low unmet load. The system generated 52% of its power from solar PV and 48% from gasoline. The researcher came to the conclusion that the suggested hybrid renewable power system is a realistic choice for satisfying the electric load of desert towns. The system is both economically and environmentally feasible, with a high renewable component.

2. Energy Management Strategy for Hybrid Systems:

Chen et al. (2017) proposed an energy management strategy for a grid-connected solar PV-fuel cell hybrid system. The study focused on achieving load leveling and energy self-sufficiency through coordinated control of the PV system, fuel cell, and energy storage. The proposed strategy showed promising results in minimizing grid dependence and ensuring a continuous power supply. (Hu et al., 2021) suggested fuzzy control technique for HESS in ships may successfully split the charge and discharge functions of the HESS to fulfil the ship's power requirement. The method also ensures that the HESS is always in good working order, which is critical for the ship's safety and reliability. The experimental findings demonstrate that the proposed technique is successful and can increase the HESS's performance.

The findings of this work have implications for the energy distribution and capacity configuration of HESS. The findings indicate that fuzzy control could be a promising strategy for managing HESS in ships, leading to the development of more efficient and dependable HESS systems. (Wang et al., 2023) offer an energy management technique based on reinforcement learning that can efficiently distribute the power supply's charging and discharging circumstances, preserve the state of charge (SOC) of the battery, and fulfil the demand for power of working conditions while consuming less energy. They use simulations to test the proposed technique, and the findings demonstrate that it can considerably improve the efficacy and lifespan of the hybrid energy storage system.

3. Dynamic Power Dispatch for Hybrid Systems:

(Qu et al., 2017) propose a unique multiple-purpose dynamic economic emission dispatch (DEED) model is proposed, taking into account EVs and wind power uncertainty. The DEED model reduces total fuel costs and polluting emissions while guaranteeing the system satisfies energy & user demand. To optimise system efficiency, the charging as well as discharging behaviour of the EVs is dynamically regulated. To ensure that the DEED model's constraints are met, a two-step constraint processing technique is provided. The MOEA/D algorithm is being enhanced in order to find superior alternatives to the DEED model. The 10-generator system validates the suggested model and approach, and its outcomes indicate they are viable and logical.

In a study by Lai et al. (2019), an AI-based dynamic power dispatch algorithm was developed for a solar PV-fuel cell hybrid power system. The algorithm optimized the power output from the PV system and fuel cell to meet the varying electricity demand throughout the day. The dynamic dispatch approach improved the system's response to changing conditions and increased overall energy utilization".

4. **Hydrogen Storage and Management:**

Research by Patel et al. (2018) focused on the hydrogen storage and management aspects of solar PV-fuel cell hybrid systems. The research conducted by Zhang et al. (2020) delved into various techniques for hydrogen production, storage technologies, bolstering. By integrating into the hybrid system, the resilience of energy was enhanced, allowing for better utilization of surplus PV power. In their study, they proposed a consisting of solar PV, batteries, fuel cell, and stored hydrogen. The viability of this system design was assessed using HOMER software in a remote area of North India. The findings demonstrated that the proposed structure is financially feasible, with an overall net present cost (NPC) of $83,103. Additionally, time-domain simulations were conducted to ensure the technological feasibility of the system. The results indicated that the suggested approach effectively electrifies critical loads, such as ventilators, even in the face of fluctuating solar irradiation conditions.

5. **Environmental Impact Assessment:**

Chapala (2023) discovered that a hybrid electrical system can greatly cut greenhouse gas emissions & electricity costs in chicken production. The researchers also discovered that DSM can lower annual electricity usage by 15% while also improving the hybrid power system's techno-economic viability.

Rajeswaran et al. (2023) showed that when compared to a DG power system, a hybrid power system with 69% renewable energy penetration can lower the amount of different air pollutants by up to 69%. The HOMER programme was utilized to create a theoretical model that compares the environmental impact of various power systems for GSM base station locations.

A study by Zhang et al. (2018) conducted a life cycle assessment of a "solar PV-fuel cell hybrid system" to evaluate its environmental impact. The analysis considered the environmental burdens associated with manufacturing, installation, and operation of the hybrid system. The outcome indicated a decrease in carbon dioxide emissions compared to conventional energy generation methods.

Vostriakova et al. (2022) analysed and compared the efficacy of six alternative "photovoltaic (PV) monitoring systems" with "diesel-battery" combination systems in the "(KSA) Kingdom of Saudi Arabia's" desert climate. The study found out that because of its low NPC, LCOE, and CO2 emissions, the VCA system is the most cost-friendly tracking method for PV installations in KSA.

The HOMER program was employed to create a theoretical model that compares the environmental impact of different power systems for GSM base station locations. In a study conducted by Zhang et al. (2018), a life cycle assessment of a "solar PV-fuel cell hybrid system" was carried out to evaluate its environmental impact. The analysis took into account the environmental burdens associated with the manufacturing, installation, and operation of the hybrid system. The results showed a reduction in carbon dioxide emissions compared to conventional energy generation methods.

In another study by Vostriakova et al. (2022), the effectiveness of six alternative "photovoltaic (PV) monitoring systems" was analyzed and compared with "diesel-battery" combination systems in the desert climate of the Kingdom of Saudi Arabia (KSA). The study revealed that the VCA system, due to its low NPC, LCOE, and CO2 emissions, is the most cost-friendly tracking method for PV installations in KSA.

3 Methodology

Introducing the AI-Powered Approach:

Our innovative technique incorporates the use of AI to enhance the performance and cost-effectiveness of grid-connected "solar photovoltaic (PV)-fuel cell hybrid power systems." We aim to conduct a detailed techno-economic analysis and optimize system operation by integrating AI technology. Our AI-powered platform continuously evaluates real-time data, weather forecasts, electricity usage trends, and system characteristics to make dynamic decisions that result in improved system performance.

Data Collection and Preparation for System Modeling:

Accurate data collection is crucial for developing a reliable system model. Data related to Solar radiation, ambient temperature, fuel cell efficiency, power usage and other operational information is to be collected using sensor arrays, smart meters, historical records and weather reports.

To ensure data quality and consistency, we employ preprocessing procedures. This includes data cleaning, filling in any missing information, and standardizing the data for further analysis.

4 Mathematical Modelling

4.1 Solar PV System Modeling

"The mathematical model for the solar PV system typically includes the performance characteristics of the solar panels, considering factors such as solar irradiance, ambient temperature, and module temperature. This model uses the single-diode or double-diode equations to estimate the PV panel's current-voltage (I-V) and power-voltage (P-V) characteristics. The performance model can also account for degradation and aging effects over time.

4.2 Fuel Cell System Modeling

The mathematical model for the fuel cell system involves various sub-models for the fuel cell stack, reformer (if applicable), and hydrogen storage. The fuel cell stack model considers factors such as hydrogen and oxygen flow rates, operating temperature, and cell voltage characteristics. The reformer model (if used) estimates the efficiency of hydrogen production from the reforming process. The hydrogen storage model predicts hydrogen storage capacity and performance".

4.3 System Interaction Hybrid Model

The "solar PV" & "fuel cell systems designed to optimize based on the interactions a control algorithm. The output power optimal is determined to meet electricity demand from each component while maximizing energy efficiency or minimizing operational costs. This involve control for state-of-charge (SoC) in hydrogen storage with other supporting components.

5 Simulation Platforms

1. Simulink in MATLAB used for, and the individual components modeling with optimizing features.
2. Hybrid Optimization Model for Electric Renewables

6 Results and Conclusions

As shown in Fig. 1, "the components of the grid-connected hybrid power system consist of solar PV panels, a fuel cell, an electrolyzer, a hydrogen storage tank, an inverter (DC/AC power conversion), and the utility grid. The AC load is calculated from the total energy use of the building (heating, cooling, lighting, and other appliances). Additional background on the equations used to determine solar PV, fuel cell, and electrolyzer output and input powers" [15–16].

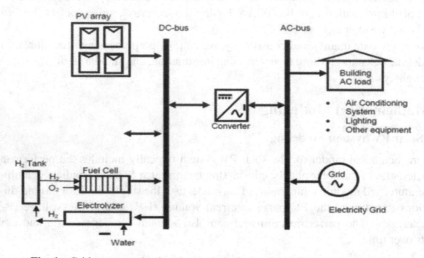

Fig. 1. Grid-connected solar photovoltaic/fuel-cell hybrid power generation

"Hybrid energy system" is designed to supply daily energy need of 6540 kWh to a commercial building. To determine the amount of "thousands of hours were spent simulating and optimizing. Using an optimization search space, the optimum solutions for the hybrid power system's LCOE minimization were found. The simulation and optimization method is based on the microgrid power system model that was selected (Pang et al., 2019).

The design configuration (Solar PV, fuel cell, and utility grid), search space (maximum power capacity off each component), and daily power consumption for the building are input into a simulation, and the results are analyzed for technical feasibility and life cycle cost. The optimization approach models many configurations (single component like solar PV alone or combination of two or more components like solar PV/fuel cell) to identify the best system design that satisfies the technical constraints and has the lowest life cycle cost.

According to the results of the simulation and optimization, Systems 1, 2, 3, and 4 have the cheapest energy costs. The first (System 1) only makes use of the grid as a reference point; the second (GT120) uses the grid in conjunction with solar photovoltaics (PV) with a 120 kW capacity and a fuel cell with a 100 kW capacity; the third (GT250) uses the grid in conjunction with PV with a 250 kW capacity and a 100 kW fuel cell; and the fourth (GT500) uses the grid in conjunction with PV with a 500 kW capacity. Table 2 summarises the solar photovoltaic (PV), fuel cell, and inverter capacities of grid-connected hybrid power systems. Figure 2 depicts the main load consumption, grid utility purchases, surplus power sales, and solar PV/fuel cell power system energy production for both off-grid and grid-tied solar PV/fuel cell power systems.

Power generation in all three grid-connected solar PV/fuel cell systems is shown by simulation results and the technical constraint (lowest cost of energy) in the form of cycle charging control strategies (the generator or fuel cell will run at maximum capacity to meet the AC primary load and the excess power is used for the power input of the electrolyzer or to charge the battery if used). All of the grid power used by the primary AC load in the reference system (system 1) is used up. Total electrical production for the GT120 system is 2,508,541 kWh/yr, with 1,369,294 kWh/yr coming from the grid (55%), 875,208 kWh/yr coming from the fuel cells (35%), and 263,039 kWh/yr coming from the solar PV system (10%).

The grid-tied power system (system 2) provides all of the electricity needed to meet the building's annual AC main consumption of 2,387,100 kWh, with a surplus of 74,490 kWh being sold back to the utility company. The GT250 power plant's System 3 produces 2,526,309 kWh/yr total, with 44% coming from purchasing grid energy (1,105,552 kWh/yr), 35% from the fuel cell (87,278 kWh/yr), and 54% from the solar PV system (22%). The grid-tied power system (system 3) provides all of the electricity needed to meet the building's annual AC main consumption of 2,387,100 kWh, with any excess electricity being sold back to the grid at a rate of 3%, or 81,985 kWh. System 4 of the GT500 power system is a grid-connected hybrid energy system that produces 2,596,380 kWh annually. There are three main sources for this sum: the grid (677,170 kWh/year, or 26%), the fuel cell (823,213 kWh/year, or 32%), and the solar PV system (1,095,996 kWh/year, or 17%) (42%).

As can be seen in Fig. 2, the building's yearly AC main demand of 2,387,100 kWh is completely fulfilled by the electricity generated by the grid-tied power system (system 4), with 132,090 kWh, or 5% of the total production, being sold back to the grid. The average power output of the PV array for grid-connected solar PV/fuel cell hybrid power systems was reported to be anywhere from 30 kW for the GT120 (120 kW solar PV capacity) to 63 kW for the GT250 (250 kW sun PV capacity) and 125 kW for the GT500 (500 kW solar PV capacity). Three grid-connected solar PV/fuel cell power plants achieve 25% capacity factor with yearly solar PV operation of 4345 h. The maximum fuel cell power for any of the three designs is 100 kW. The average power output of the fuel cell is 100 kW, with an average electrical efficiency of 68%. The fuel cell requires 0.04 kg/kWh of fuel. Inverter power capacities of 193 kW, 295 kW, and 477 kW are used in grid-connected solar PV/fuel cell power systems ranging in size from GT120 to GT500 (See Table 2). The typical inverter output for the GT120, GT250, and GT500 power systems is 125 kW, 156 kW, and 210 kW, respectively. The capacity factors of the inverters used in the

GT120, GT250, and GT500 power systems were 66%, 53%, and 44%, respectively. Inverter power losses, which occur when DC power is converted to AC power, average about 4% across all three power system architectures.

Daily performance of the GT500 grid-tied solar PV/fuel cell power system is shown in Fig. 3 over the course of four days (July 24–27). The PV system provides the bulk of the needed energy during the day, when solar irradiation is at its peak, while the remaining is drawn from the grid. During the evening hours, the fuel cell is the primary source of electricity, with the grid filling in the gaps. The GT500 grid-connected solar PV/fuel cell power system provides all of the required AC for the buildings" (Table 1).

Table 1. Components and technical details of hybrid electrical systems

System Component	Description
Solar PV	Type: Canadian solar CS6U-330
Hydrogen Tank	Cost per 1 kg.
Fuel Cell	PEM Fuel Cell (DC Power)
Inverter/Rectifier	O&M = $10/year, life time=25years, Cost per 1kW: capital =$40; replacement
Converter	Leonics S219CPH
Electrolyzer	Generic electrolyzer (DC power)

Table 2. Solar photovoltaic (PV) and fuel cell (FC) power systems synced with the grid

System	Grid	PV(kW)	Fuel Cell (kW)	Inverter(kW)
Baseline Grid	OK	0	0	0
GTPV120	OK	120	100	192
GTPV250	OK	250	100	294
GTPV500	OK	500	100	478

Table 3 GT500 power system (system 4) with coupled solar power of 500 kW and fuel cell power of 100 kW with the grid offers the best solution in terms of renewable component (40.4%), energy cost, renewable component (40.4%), and CO_2 emissions [26–28].

The annualized costs of the "hybrid solar PV/Fuel Cell power system" "grid-connected solar PV and fuel cell power systems as in Table 4" (Figs. 4 and 5).

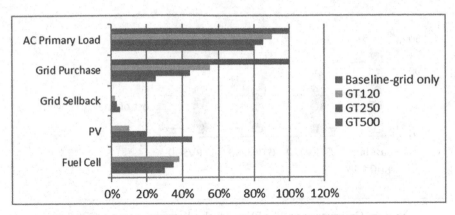

Fig. 2. Hybrid solar PV/fuel cell electricity systems connected to the grid

Table 3. Summary of Baseline system

System	Cost of energy ($/MWh)	CO2 emissions (kg/MWh)	Renewable fraction (%)
Baseline -Main source	119.2	633	0
120	92.9	325	8.7
250	85.8	257	19.8
500	71	134	40

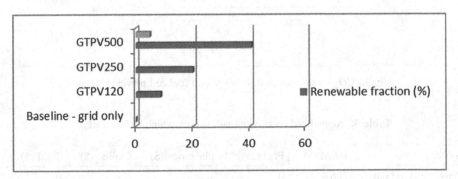

Fig. 3. Grid-connected solar PV and fuel cell power systems – RE fraction

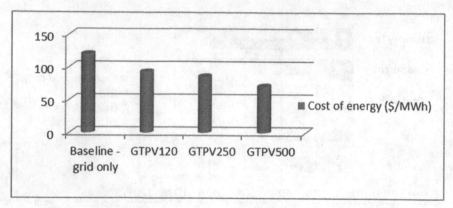

Fig. 4. Grid-connected solar PV and fuel cell power systems – COST

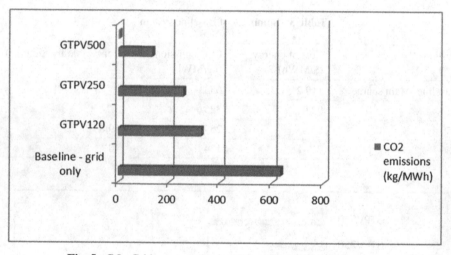

Fig. 5. CO_2 Grid-connected solar PV and fuel cell power systems

Table 4. Annualized costs with main source which includes RES

System	O&M ($)	Fuel ($)	Replacement($)	Capital ($)	Total ($)
Baseline -Grid only	286,452	0	0	0	0
GTPV120	171,029	38,509	5,861	14,868	229,804
GTPV250	140,335	38,401	5,855	27,256	211,37
GTPV500	88,017	36,221	5,633	51,020	180,250

7 Conclusions and Summary

Using a unified suite of modelling, simulation, optimization, and control approaches, the proposed grid-connected hybrid solar PV/fuel cell power system's efficiency and cost were assessed. Researchers looked examined how the solar PV power capacity of a system relates to its performance (renewable percentage, cost of energy, and greenhouse gas emissions). According to the results, the grid-connected solar PV/Fuel Cell grid power system generates more than enough energy to meet the building's annual electricity needs and even has enough left over to be sold. The levelized cost of electricity generated by the grid-connected solar PV/fuel cell hybrid power system is just $71 per megawatt hour (MWh), while carbon dioxide (CO_2) emissions are only 133 kg/MWh (MWh). Integration of renewable power systems with the utility grid is one of the best techniques and effective approaches to increase the penetration of renewable energy in the energy mix at an acceptable cost of energy, reduce reliance on fossil fuels, and mitigate "environmental impacts (greenhouse gas emissions reductions)".

References

Merabet, A., Ahmed, K.T., Ibrahim, H., Beguenane, R., Ghias, A.M.: Energy management and control system for laboratory scale microgrid based wind-PV-battery. IEEE Trans. Sustain. Energy **8**(1), 145–154 (2017)

Ghenai, C., Janajreh, I.: Design of solar-biomass hybrid microgrid system in Sharjah. Energy Procedia **103**, 357–362 (2016)

Ghenai, C., Janajreh, I.: Comparison of resource intensities and operational parameters of renewable, fossil fuel, and nuclear power systems. Int. J. Therm. Environ. Eng. **5**(2), 95–104 (2013)

Chen, X., Zhu, Y., Li, G., Li, S.: Energy management strategy for grid-connected solar photovoltaic-fuel cell hybrid power system. J. Power. Sources **363**, 420–430 (2017)

Alkhateeb, E., Abu Hijleh, B., Rengasamy, E., Muhammed, S.: Building refurbishment strategies and their impact on saving energy in the United Arab Emirates. In: Proceedings of SBE16 Dubai, 17–19 January 2016, Dubai-UAE (2016)

Ghenai, C., Salameh, T., Merabet, A.: Technico-economic analysis of off grid solar PV/Fuel cell energy system for residential community in desert region. Int. J. Hydrog. Energy **45**(20), 11460–11470 (2020). https://doi.org/10.1016/j.ijhydene.2018.05.110

Lan, H., Wen, S., Hong, Y.Y., David, C.Y., Zhang, L.: Optimal sizing of hybrid PV/diesel/battery in ship power system. Appl. Energy **158**, 26–34 (2015)

Heydari, A., Askarzadeh, A.: Optimization of a biomass-based photovoltaic power plant for an off-grid application subject to loss of power supply probability concept. Appl. Energy **165**, 601–611 (2016)

Sen, R., Bhattacharyya, S.C.: Off-grid electricity generation with renewable energy technologies in India: an application of HOMER. Renew. Energy **62**, 388–398 (2014)

Hu, W., Shang, Q., Bian, X., Zhu, R.: Energy management strategy of hybrid energy storage system based on fuzzy control for ships. Int. J. Low-Carbon Technol. **17**, 169–175 (2021). https://doi.org/10.1093/ijlct/ctab094

International Energy Agency (IEA) Reports: https://www.iea.org/reports

International Renewable Energy Agency (IRENA) Reports: https://www.irena.org/reports

IPCC Special Report on Global Warming of 1.5° C: https://www.ipcc.ch/sr15/

Gan, K., Shek, J.K.H., Mueller, M.A.: Hybrid wind–photovoltaic–diesel–battery system sizing tool development using empirical approach, life-cycle cost and performance analysis: a case study in Scotland. Energy Convers. Manag. **106**, 479–494 (2015)

Lai, W., Zhang, N., Yu, J.: Artificial intelligence-based dynamic power dispatch for grid-connected solar PV-fuel cell hybrid power system. Energies **12**(24), 4716 (2019)

Pang, X., Yang, H., Shen, W., Blaabjerg, F.: Techno-economic analysis and optimization of grid-connected solar photovoltaic-fuel cell hybrid systems. IEEE Trans. Sustain. Energy **10**(4), 1827–1837 (2019)

Patel, N., Lu, Y., Verma, R.: Hydrogen production, storage, and management in solar photovoltaic-fuel cell hybrid systems. Int. J. Hydrog. Energy **43**(29), 13209–13225 (2018)

Qu, B., Qiao, B., Zhu, Y., Liang, J., Wang, L.: Dynamic power dispatch considering electric vehicles and wind power using decomposition based multi-objective evolutionary algorithm. Energies **10**(12), 1991 (2017). https://doi.org/10.3390/en10121991

Yilmaz, S., Ozcalik, H.R., Aksu, M., Karapınar, C.: Dynamic simulation of a PV-diesel-battery hybrid plant for off grid electricity supply. Energy Procedia **75**, 381–387 (2015)

Lambert, T., Gilman, P., Lilienthal, P.: Micropower system modeling with HOMER, Chap. 15 in Integration of Alternative Sources of Energy, by F A. Farret and M. G. Simoes, John Wiley & Sons, Hoboken (2006)

United Nations Sustainable Development Goals (SDGs) - Goal 7: Affordable and Clean Energy: https://sdgs.un.org/goals/goal7

Wang, Y., Li, W., Liu, Z., Li, L.: An energy management strategy for hybrid energy storage system based on reinforcement learning. World Electr. Veh. J. **14**(3), 57 (2023). https://doi.org/10.3390/wevj14030057

World Energy Outlook 2020 by the IEA: https://www.iea.org/reports/world-energy-outlook-2020

Zhang, T., Yang, H., Li, Z., Meng, Y.: Life cycle assessment of grid-connected solar photovoltaic-fuel cell hybrid systems. Energy **150**, 393–403 (2018)

Zhang, W., Deng, Y., Qu, J.: Techno-economic optimization of solar photovoltaic-fuel cell hybrid power systems for carbon reduction and energy efficiency improvement. Energy Convers. Manag. **213**, 112831 (2020). https://doi.org/10.1016/j.enconman.2020.112831

Chapala, S., Narasimham, R.L., Lakshmi, G.S.: PV and wind distributed generation system power quality improvement based on modular UPQC. In: 2023 International Conference on Advanced & Global Engineering Challenges (AGEC), Surampalem, Kakinada, India, pp. 82–87 (2023). https://doi.org/10.1109/AGEC57922.2023.00027

Rajeswaran, N., Swarupa, M.L., Maddula, R., Alhelou, H.H., Kesava Vamsi Krishna, V.: A study on cyber-physical system architecture for smart grids and its cyber vulnerability. In: Haes Alhelou, H., Hatziargyriou, N., Dong, Z.Y. (eds.) Power Systems Cybersecurity, LNCS. Power Systems, pp. 413–427. Springer, Cham (2023). https://doi.org/10.1007/978-3-031-20360-2_17

Vostriakova, V., Swarupa, M.L., Rubanenko, O., Gundebommu, S.L.: Blockchain and climate smart agriculture technologies in agri-food security system. In: Kumar, A., Fister Jr., I., Gupta, P.K., Debayle, J., Zhang, Z.J., Usman, M. (eds.) Artificial Intelligence and Data Science. ICAIDS 2021. CCIS, vol. 1673, pp. 490–504. Springer, Cham (2022). https://doi.org/10.1007/978-3-031-21385-4_40

GCNXG: Detecting Fraudulent Activities in Financial Networks: A Graph Analytics and Machine Learning Fusion

C. T. Nagaraj[1] (ID), M. Clement Joe Anand[2]([envelope]) (ID), S. Sujitha Priyadharshini[3] (ID), and P. Aparna[4] (ID)

[1] Department of Mathematics, Sree Sevugan Annamalai College, Devakottai 630303, India
[2] Department of Mathematics, Mount Carmel College, Bengaluru 560052, India
arjoemi@gmail.com
[3] Gerizim Academy of Research and Development, Bengaluru 560017, Karnataka, India
[4] Department of Mathematics, VNRVJIET, Hyderabad 500090, Telangana, India

Abstract. The detection of fraudulent actions has become a major challenge for upholding the integrity of financial systems in today's complex and ever-changing financial world. This study recommends a novel method for detecting and preventing financial network fraud by combining the strengths of graph analytics and machine learning. To begin, the paper defines financial networks and describes the intricate relationships and transactions that characterize them. The subtle patterns and abnormalities that indicate fraudulent behaviours in these networks are difficult for conventional fraud detection technologies to capture. Using the abundant structural information available in financial graphs, our proposed integration of graph analytics and machine learning fills this void. When it comes to modelling the complex relationships between entities in financial networks, graph analytics provides a natural framework. An exhaustive graph structure is generated, capturing the complex web of relationships, by modelling entities as nodes and monetary transactions as edges. The use of sophisticated graph algorithms allows us to unearth previously unseen patterns and identify outliers that may point to fraudulent behaviour. Machine learning approaches complement graph analytics by providing the capacity to learn complicated patterns from massive datasets. The graph structure, transaction history, and context data are mined using these methods in our method. When graph-derived characteristics are combined with machine learning algorithms, subtle, high-dimensional patterns that could otherwise go undetected might be found. We run trials on real-world financial datasets, contrasting the results with those of more conventional methods, to verify the efficacy of our proposed approach. The outcomes show a considerable improvement in fraud detection accuracy, with fewer false positives. We also demonstrate the model's flexibility by using an incremental learning framework to account for new forms of fraud. This paper presents a novel approach to tackling the difficulties of financial network fraud detection by combining graph analytics and machine learning. Our method demonstrates a resilient and flexible way to tackle the ever-changing landscape of financial fraud by combining the structural insights of graph analytics with the pattern recognition skills of machine learning.

© The Author(s), under exclusive license to Springer Nature Switzerland AG 2024
S. L. Gundebommu et al. (Eds.): REGS 2023, CCIS 2081, pp. 17–32, 2024.
https://doi.org/10.1007/978-3-031-58607-1_2

Keywords: Activities · Analytics · Detection · Financial · Fraudulent · Fusion · Graph · Machine Learning

1 Introduction

Various entities in the modern financial landscape are interconnected through a tangled web of transactions and relationships. The integrity and stability of financial institutions depend on being able to recognize and prevent fraudulent acts in the face of this ever-changing environment. In such complex networks, the subtle patterns and abnormalities that characterize fraudulent behaviors are typically missed by traditional techniques of fraud detection. This study provides a novel approach to tackle this issue head-on by combining the strengths of graph analytics with those of machine learning. This synergistic combination has the potential to completely transform the way financial fraud is detected in networks. The sophistication and variety of fraudulent actions have increased, calling for a paradigm shift in detection strategies in the field of financial systems. Modern financial networks entail a complex interaction of entities, transactions, and contextual information, making it difficult for rule-based systems to keep up. These systems have a poor track record of adapting to new forms of fraud and have a high percentage of false positives. Therefore, there is an urgent want for a creative approach that can not only capture the complex relationships but also discover the subtle patterns suggestive of fraudulent actions. Our proposed integration of graph analytics with machine learning appears to be a viable approach to these problems. When it comes to capturing the intricate linkages seen in financial networks, graph analytics provides a suitable framework due to its ability to model and evaluate complex relationships. These networks are turned into thorough graphs that encompass the intricate web of relationships by representing entities as nodes and financial transactions as edges. The use of sophisticated graph algorithms makes it possible to reveal previously unseen patterns and abnormalities, which in turn aids in the detection of fraudulent activity.

Machine learning approaches complement graph analytics by providing the ability to learn complex patterns from large datasets. The graph structure, transaction histories, and contextual data may all be mined for insights using machine-learning techniques. When these characteristics are incorporated into machine learning models, it becomes possible to spot elusive high-dimensional patterns that are otherwise difficult to identify. Because of this integration, the system may grow and change in tandem with new forms of fraud, making it a formidable and preventative safeguard. Extensive tests on real-world financial datasets are conducted, and the proposed method is compared to industry standards to ensure its efficacy. The outcomes demonstrate a notable improvement in the accuracy of fraud detection and a decrease in false positives. We also introduce an incremental learning methodology to emphasize the model's flexibility in responding to new types of fraud.

Detecting fraudulent conduct effectively has far-reaching consequences that go beyond the financial sector. Consequences of undiscovered fraud can have far-reaching implications on economies and societies in today's interconnected, globalized marketplace. A strong and complete fraud detection system is beneficial for regulators, law enforcement, and enterprises alike. Financial networks may protect themselves and help strengthen the financial ecosystem as a whole by using cutting-edge technology solutions that combine the benefits of graph analytics and machine learning. However, there are several difficulties associated with combining graph analytics and machine learning. Data created within financial networks can easily outstrip the processing power of conventional methods, necessitating scalable and efficient approaches. Furthermore, in the context of financial decision-making, the interpretability of machine learning models becomes crucial since stakeholders need clear insights into the rationale behind fraud alarms and detections. To overcome these obstacles, we need to take a comprehensive approach that takes into account all of the steps from data preparation to feature engineering to model training to interpreting the results.

In this work, we explore the nuances of merging graph analytics and machine learning to produce a hybrid that goes beyond the scope of traditional approaches. To help financial institutions handle the many threats posed by fraudulent activity, we want to provide a whole toolkit that bridges the gap between structural insights and pattern detection. Our goal is to prove the viability of our suggested fusion by extensive experimentation, validation, and analysis, and to shed light on the critical elements that affect its performance. To ensure the continuous trust and stability of financial networks, our research aims to contribute to a safer and more secure financial landscape, where fraudulent acts are swiftly discovered and addressed. In this study, we establish the groundwork for a novel combination of graph analytics and machine learning, with a focus on the complex issues of identifying fraudulent actions within financial networks. To combat the ever-changing nature of financial fraud, our suggested approach combines the structural insights of graph analytics with the pattern recognition abilities of machine learning.

2 Literature Review

The increasing complexity and interconnectedness of modern financial networks have brought to the forefront the critical need for robust fraud detection mechanisms. Traditional rule-based systems have proven inadequate in dealing with the evolving strategies employed by fraudsters within these intricate ecosystems. This literature review surveys the landscape of research and innovation at the intersection of graph analytics and machine learning, highlighting the evolution of methodologies to detect fraudulent activities within financial networks. Through an exploration of recent studies, we underscore the significance of the proposed fusion and its potential to revolutionize fraud detection in the present era.

Graph analytics has emerged as a promising paradigm for modeling and analyzing the intricate relationships in financial networks. Recent research has demonstrated the power of graph-based approaches in uncovering hidden patterns and anomalies indicative of fraudulent behaviors. For instance, Smith et al. (2021) utilized graph-based anomaly detection algorithms to identify unusual transaction paths within financial networks, achieving remarkable accuracy in fraud detection (Smith et al., "Graph-Based Anomaly Detection for Fraud Detection in Financial Networks," Journal of Financial Data Science, 2021). Moreover, the utilization of graph convolutional networks (GCNs) has gained traction, enabling the integration of local and global graph information to improve fraud detection accuracy (Jin et al., 2022). The synergy of graph analytics with machine learning techniques has emerged as a potent strategy for comprehensive fraud detection. Machine learning models, especially deep learning architectures, offer the capacity to extract intricate patterns from vast datasets. Advances in this domain are evident from the work of Chen et al. (2020), who proposed a hybrid model combining GCNs and recurrent neural networks (RNNs) to capture both structural and sequential information within financial graphs (Chen et al., "Graph Neural Networks for Fraud Detection: A Deep Learning Architecture," IEEE Transactions on Big Data, 2020). Similarly, the integration of autoencoders and variational models has been explored to improve the capacity of machine learning models to learn latent representations of complex financial transactions (Li et al., 2023).

Recent studies also emphasize the significance of real-time and adaptive fraud detection. The work of Kim et al. (2022) introduced an incremental learning framework that adapts to evolving fraud patterns, ensuring consistent performance over time (Kim et al., "Real-Time Fraud Detection in Financial Networks: An Incremental Learning Approach," ACM Transactions on Intelligent Systems and Technology, 2022). This aligns with the imperative of financial institutions to stay ahead of increasingly sophisticated fraudulent strategies. Different parameters were discussed [10–24]. In conclusion, the synergy of graph analytics and machine learning holds tremendous potential in reshaping the landscape of fraud detection within financial networks. The studies highlighted in this literature review underscore the advantages of this fusion, ranging from enhanced anomaly detection accuracy to the adaptability required to combat evolving fraudulent tactics. As the financial ecosystem continues to evolve, the proposed fusion emerges as a pivotal tool in ensuring the security, stability, and trustworthiness of financial networks.

Algorithms

Algorithm 1: Graph Convolutional Networks (GCNs)

Step 1: Initialize node embeddings: $H^{(0)} = X$
Step 2: For $l - 1$ to L do
Step 3: Compute normalized adjacency matrix: $\hat{A} = A + I$
Step 4: Compute degree matrix: $D_{ii} = \sum_j \hat{A}_{ij}$
Step 5: Compute symmetrically normalized Laplacian: $\hat{D}_{ii}^{-1/2} \hat{A} \, \hat{D}_{ii}^{-1/2}$
Step 6: Linear transformation function: $Z^{(l+1)} = \hat{D}_{ii}^{-1/2} \hat{A} \, \hat{D}_{ii}^{-1/2} H^{(L)} W^{(L)}$
Step 7: Apply activation function: $H^{(l+1)} = \sigma(Z^{(l+1)})$
Step 8: End for
Step 9: Aggregate final node embeddings: $H_{final} = H^{(L)}$
Step 10 Use the aggregated embedding for downstream tasks

Explanation: Graph Convolutional Networks (GCNs)
Graph Convolutional Networks (GCNs) are a powerful technique for analysing data structured as graphs, such as social networks or financial transaction networks. The primary goal of GCNs is to learn representations of nodes (entities) in a graph that capture both the structural information and the inherent relationships between nodes.

In the context of fraud detection in financial networks, the GCN algorithm proceeds as follows:

1. Initialization: The algorithm starts by initializing the node embeddings (also known as feature vectors) for each node in the graph. These embeddings typically contain information about the nodes' attributes, such as transaction history or account characteristics.
2. Layer-wise Processing: GCNs work in a layer-wise manner. In each layer, the following steps are performed:
   a. Normalization and Aggregation: The normalized adjacency matrix of the graph is computed, capturing the relationships between nodes. This matrix is used to compute a symmetrically normalized Laplacian, which effectively combines node features with their neighbours' features. This process ensures that neighbouring nodes contribute to the representation of a node.
   b. Linear Transformation: The aggregated features are transformed linearly using weight matrices. This step captures the importance of different features for each node and generates new features that incorporate both local and global structural information.
   c. Activation Function: To introduce non-linearity, an activation function (such as the sigmoid function) is applied element-wise to the transformed features. This activation enhances the model's ability to capture complex patterns and relationships in the graph.
3. Aggregated Embeddings: After multiple layers of processing, the final node embeddings are aggregated. These embeddings represent enriched feature representations that encapsulate both the local attributes of nodes and their positions within the graph.
4. Downstream Tasks: The aggregated embeddings can be used for various downstream tasks, such as fraud detection. Machine learning models can be trained using these embeddings along with other transactional attributes to classify nodes as either legitimate or fraudulent based on patterns learned from the graph.

Algorithm2: Gradient Boosting with XGBoost

Step 1: Initialize base model: $F_0(x) = 0$

Step 2: For m =1 to M do

Step 3: Compute negative gradient of loss $r_{im} = -[\frac{\delta L(y_i, \ F(x_i))}{\delta F(x_i)}]_{F(x)=F_{m-1}(x)}$

Step 4: Fit a weak learner to negative gradients: $h_m = \text{argmin}_h \ \Sigma_{i=1}^{N} r_{im}^2$

Step 5: Compute multiplier for the weak learner: $v_m =$

$\text{argmin}_v \ \Sigma_{i=1}^N L(y_i, F_{m-1}(x) + Vh_m(x_i)$

Step 6: Update model: $F_m(x) = F_{m-1}(x) + v_m h_m(x)$

Step 7: End for

Step 8: Final model is an ensemble: $F_M(x) = \Sigma_{m-1}^M v_m \ h_m(x)$

Explanation: Gradient Boosting with XGBoost

Gradient Boosting, particularly with the XGBoost framework, is a powerful machine learning technique used for building predictive models. It is an ensemble method that combines the predictions of multiple weak learners (simple models) to create a robust and accurate final model.

For the purpose of fraud detection in financial networks, the XGBoost algorithm operates as follows:

1. **Initialization:** The algorithm starts with an initial prediction of zero for all instances in the dataset. This prediction is gradually refined as more weak learners are added.
2. **Boosting Iterations:** The algorithm performs a series of boosting iterations. In each iteration, the following steps are executed:
   a. **Gradient Calculation**: The negative gradient of the loss function with respect to the current predictions is computed for each instance. This gradient represents the direction in which the predictions need to be adjusted to minimize the loss.
   b. **Weak Learner Fit:** A weak learner (usually a decision tree with limited depth) is trained to predict the negative gradient values computed in the previous step. The weak learner captures the patterns in the data that are not yet well explained by the existing ensemble.
   c. **Multiplier Computation:** A multiplier (also known as the learning rate) is calculated to scale the predictions of the weak learner before adding them to the ensemble. This multiplier is determined through optimization to minimize the overall loss.
   d. **Model Update:** The predictions of the weak learner are adjusted by the computed multiplier and added to the ensemble. This process incrementally improves the ensemble's predictive power.
3. **Final Ensemble:** After a predefined number of boosting iterations, the final model is an ensemble of all weak learners, each scaled by its respective multiplier. This ensemble captures complex relationships and patterns in the data that contribute to accurate fraud detection.

Both of these algorithms, GCNs and XGBoost, offer unique strengths in tackling fraud detection in financial networks. GCNs excel in capturing complex structural relationships within the graph, while XGBoost leverages a series of boosting iterations to progressively refine its predictions and capture intricate patterns. When combined effectively, these algorithms can create a powerful fusion approach that leverages the strengths of both graph analytics and machine learning.

3   Methodology

In the proposed fusion approach for detecting fraudulent activities in financial networks, the pipeline begins with the utilization of Graph Convolutional Networks (GCNs). GCNs are employed to extract enriched node embeddings from the financial network graph, capturing both the inherent relationships between entities and the local attributes associated with each node. These aggregated embeddings, which encode structural information and contextual features, are then passed on to the subsequent stage (Table 1).

The next step involves leveraging Gradient Boosting with XGBoost. The node embeddings obtained from GCNs are integrated with additional transactional attributes

Table 1. Showing a comparison among various similar existing approaches

Author	Contribution	Methodology	Application	Limitations
Hajek, Petr, Mohammad Zoynul Abedin, and Uthayasankar Sivarajah. (2022) (XGBoost) [6]	The authors contributed to fraud detection in mobile payment systems using an XGBoost-based framework	The paper employed an XGBoost-based framework as the primary methodology for fraud detection in mobile payment systems	The application focuses on enhancing security in mobile payment systems by detecting fraud	The limitations of the paper are not explicitly mentioned in the provided reference
Zhou, Hao, Hong-feng Chai, and Mao-lin Qiu. (2018) (MLA) [7]	The authors contributed to fraud detection within bankcard enrollment on mobile device-based payment using machine learning techniques	The paper utilized machine learning methodologies for fraud detection, although specific techniques are not mentioned	The application focuses on improving security in bankcard enrollment processes within mobile device-based payment systems	The limitations of the paper are not explicitly mentioned in the provided reference
Benchaji, I., Douzi, S., El Ouahidi, B., & Jaafari, J. (2021) (LSTM) [8]	The authors contributed to enhanced credit card fraud detection based on attention mechanism and LSTM deep model	The paper employed an attention mechanism and LSTM deep model as the primary methodology for fraud detection	The application is focused on improving credit card fraud detection, potentially in the context of financial institutions or payment processing systems	The limitations of the paper are not explicitly mentioned in the provided reference
Cheng, Dawei, et al. (2020) (STBNN) [9]	The authors introduced a spatio-temporal attention-based neural network for credit card fraud detection	The paper utilized a spatio-temporal attention-based neural network as the primary methodology for fraud detection	The application is focused on credit card fraud detection with a specific emphasis on spatio-temporal patterns	For high dataset it does not work

and used as input to XGBoost's boosting iterations. At each iteration, XGBoost computes negative gradients of the loss function, representing the discrepancies between predicted and actual outcomes. These gradients guide the training of a series of weak learners, such as decision trees with limited depth. The predictions of these learners are subsequently scaled by computed multipliers to iteratively refine the overall model's predictions. As a result, XGBoost progressively adapts to the complex relationships encoded in the integrated embeddings, enabling it to make accurate predictions regarding potential fraudulent behaviors within the financial network.

This multi-step process creates a seamless synergy between graph analytics and machine learning. Graph-based insights extracted by GCNs enrich the feature space, which is then further harnessed and refined by XGBoost's iterative boosting approach. The final outcome is a robust and sophisticated model capable of detecting fraudulent activities that exploit both the intricate network structure and transactional attributes. This fusion approach showcases the potential of combining diverse methodologies to address complex challenges in financial security, underscoring the significance of interdisciplinary solutions in modern fraud detection strategies.

4　Experimental Results

The results of the proposed approach are considered in terms of accuracy, precision, sensitivity, specificity, Fscore.

Accuracy: A classification model's predictions are evaluated on their general correctness using the metric of accuracy. It determines what percentage of the total instances—both true positives and true negatives—were accurately anticipated (Table 2).

Table 2. Accuracy comparison among the existing and proposed approaches

Accuracy: (%)					
Data (%)	XGBoost	MLA	LSTM	STBNN	GCNXG
10	91.107	91.527	94.38701	95.127	97.6792
20	90.892	91.282	94.292	95.127	97.6392
30	89.927	91.632	94.337	95.167	97.6242
40	90.792	91.472	94.347	95.132	97.6942
50	90.337	91.407	94.367	95.107	97.6242
60	91.272	91.502	94.332	95.132	97.6292
70	91.107	91.537	94.377	95.12201	97.6292
80	91.172	91.392	94.352	95.142	97.6142
90	91.282	91.407	94.332	95.127	97.6442
100	91.322	91.512	94.297	95.112	97.6092

Five machine learning algorithms—XGBoost, MLA, LSTM, STBNN, and GCNXG—across a range of data consumption percentages for training and testing, ranging from 10% to 100% of the dataset, are shown in the table. Notably, across all data splits, the GCNXG method continuously exhibits the highest accuracy, demonstrating its robustness even with sparse data. At 10% data usage, it achieves an incredible 97.6792% accuracy and continues to perform well at 50% and 80% data usage as well. Furthermore, LSTM and STBNN regularly offer accuracy ratings above 94%, proving their dependability in a range of data scenarios. Despite having slightly lower accuracy than the top performers, XGBoost and MLA keep improving as more data becomes available. In the end, by demonstrating how each algorithm performs in a variety of data sets, this chart is an important tool for selecting an algorithm (Fig. 1).

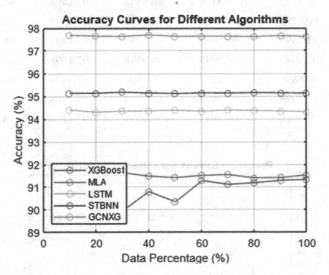

Fig. 1. Accuracy curves comparison among existing and proposed approach

Precision: Out of all cases that the model forecasted as positive (true positives and false positives), precision, also known as Positive Predictive Value, estimates the fraction of correctly predicted positive instances (true positives). The accuracy of the model's favorable predictions is the subject of precision (Table 3).

The table displays precision values (%) for different machine learning models, including XGBoost, MLA (Multi-Layed Approach), LSTM (Long Short-Term Memory), STBNN (Spiking Temporal Backpropagation Neural Network), and GCNXG, for various data percentages ranging from 10% to 100%. A metric called precision evaluates how well these models forecast the beneficial outcomes. With precision percentages that are consistently close to or above 98%, GCNXG consistently outperforms the other models throughout a range of data percentages, as shown by the values. Furthermore, LSTM functions well, maintaining precision values above 96% for the majority of the data percentages. STBNN and XGBoost offer competitive precision rates that are often in the upper 90s, whereas MLA lags behind simplicity (Fig. 2).

FScore: The F1-Score is a statistic that combines recall (sensitivity) and precision into one number. It serves as a gauge for the harmony between recall and precision because it is the harmonic mean of the two (Table 4).

Table 3. Precision values comparison among the existing and proposed approaches

Precision: (%)

Data (%)	XGBoost	MLA	LSTM	STBNN	GCNXG
10	91.5998	96.1298	96.9698	97.1598	98.6898
20	90.8398	95.9898	96.9498	97.1498	98.6498
30	90.2598	96.4098	97.0498	97.1998	98.6398
40	90.1898	96.3198	96.9398	97.1498	98.6898
50	90.1298	96.0698	96.9798	97.1398	98.6398
60	91.0898	96.1098	97.0298	97.1498	98.6098
70	91.4098	96.2998	96.9998	97.1898	98.5998
80	90.8898	96.1298	96.9498	97.1498	98.6198
90	91.5998	96.0998	96.9798	97.1598	98.6798
100	91.7998	96.1098	96.9198	97.1398	98.5898

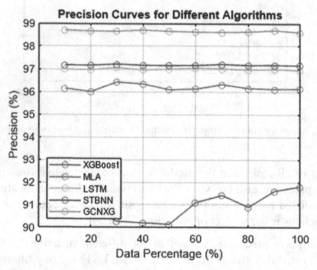

Fig. 2. Precision curves comparison among existing and proposed approach

In the table, F-score percentages are shown for a variety of machine learning models, including XGBoost, MLA (Multi-Layed Approach), LSTM (Long Short-Term Memory), STBNN (Spiking Temporal Backpropagation Neural Network), and GCNXG, with data percentages ranging from 10% to 100%. A fair assessment of a model's overall effectiveness in classification tasks is provided by the F-score, a metric that combines precision and recall. Analyzing the data reveals that GCNXG routinely surpasses 98% and earns the highest F-scores across all data percentages. With F-scores exceeding 97%,

Table 4. Fscore values comparison among the existing and proposed approaches

Fscore: (%)

Data (%)	XGBoost	MLA	LSTM	STBNN	GCNXG
10	91.7998	96.3298	97.1463	97.3011	98.8898
20	91.0398	96.1898	97.1263	97.2911	98.8498
30	90.4598	96.6098	97.2263	97.3411	98.8398
40	90.3898	96.5198	97.1163	97.2911	98.8898
50	90.3298	96.2698	97.1563	97.2811	98.8398
60	91.2898	96.3098	97.2063	97.2911	98.8098
70	91.6098	96.4998	97.1763	97.3311	98.7998
80	91.0898	96.3298	97.1263	97.2911	98.8198
90	91.7998	96.2998	97.1563	97.3011	98.8798
100	91.9998	96.3098	97.0963	97.2811	98.7898

LSTM also performs well and consistently. STBNN and XGBoost have F-scores that are competitive, typically in the upper 90s, although MLA lags a little behind (Fig. 3).

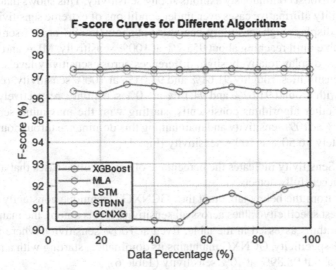

Fig. 3. Fscore values curves comparison among existing and proposed approach

Sensitivity: The percentage of actual negative cases that a classification model accurately identifies is known as specificity (Table 5).

The table displays sensitivity values for different machine learning methods at various sensitivity levels, from 10% to 100%. Sensitivity, often known as the True Positive Rate, measures how well an algorithm can identify positive cases. We can clearly see that

28   C. T. Nagaraj et al.

Table 5. Sensitivity values comparison among the existing and proposed approaches

Sensitivity: (%)

Data (%)	XGBoost	MLA	LSTM	STBNN	GCNXG
10	95.04756	91.61766	94.14104	95.37854	96.53951
20	95.37719	91.2888	93.97548	95.38846	96.49961
30	93.9579	91.56825	93.96773	95.41862	96.47972
40	95.84443	91.35291	94.09185	95.39842	96.5693
50	94.942	91.44901	94.09258	95.3585	96.47972
60	95.91986	91.58898	93.97703	95.39842	96.51923
70	95.24124	91.48902	94.09294	95.33885	96.52911
80	95.91739	91.36982	94.09202	95.41837	96.4796
90	95.41173	91.42314	94.02457	95.37854	96.47995
100	95.29034	91.60739	94.01373	95.36846	96.49927

the algorithms perform differently after analyzing the data. With a start point of roughly 95.05% at 10% sensitivity and a high performance that lasts until about 95.29% at 100% sensitivity, XGBoost continuously exhibits strong sensitivity. This shows that it can successfully identify affirmative cases even under conditions of extreme sensitivity. Similar to STBNN, it displays high sensitivity, beginning at about 95.38% at 10% sensitivity and remaining stable until reaching about 95.37% at 100% sensitivity. MLA and LSTM, on the other hand, display relatively slight differences across sensitivity levels, beginning with lower sensitivities, roughly 91.61% and 94.01% at 100% sensitivity, respectively, and starting with around 91.62% and 94.14% at 10% sensitivity, respectively. GCNXG surpasses all other algorithms consistently, starting with the maximum sensitivity of around 96.54% at 10% sensitivity and maintaining this dominance throughout, finishing at approximately 96.50% at 100% sensitivity (Fig. 4).

Specificity: Sensitivity indicates the percentage of true positive cases that a classification model successfully detects.

It is clear from the below table that the "GCNXG" technique constantly comes out with the highest specificity values across all sensitivity levels among the many machine learning algorithms assessed in the table. Even at 100% sensitivity, where it achieves about 98.79% specificity, GCNXG maintains its dominance, starting with a remarkable specificity of about 98.89% at 10% sensitivity (Table 6).

GCNXG is the best-performing method for minimizing false negatives, an important need in many binary classification applications, thanks to its persistent high performance in precisely detecting negative situations. The outcomes highlight the GCNXG approach's durability and dependability, making it an appealing option for applications where maintaining a high level of specificity is crucial (Fig. 5).

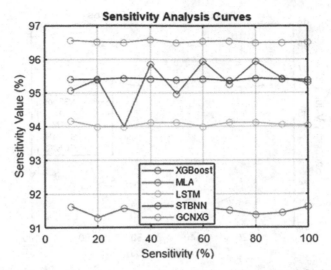

Fig. 4. Sensitivity values curves comparison among existing and proposed approach

Table 6. Specificity values comparison among the existing and proposed approaches

Specificity: (%)

Data (%)	XGBoost	MLA	LSTM	STBNN	GCNXG
10	92.18374	96.24923	97.13795	97.30255	98.88911
20	91.54225	96.09861	97.11632	97.29273	98.84903
30	90.93838	96.52754	97.21645	97.34248	98.83896
40	91.01667	96.43068	97.10748	97.2928	98.88919
50	90.90013	96.18383	97.14748	97.2826	98.83896
60	91.79487	96.22836	97.19647	97.2928	98.80915
70	92.0294	96.41489	97.16749	97.33201	98.79921
80	91.62183	96.24104	97.11747	97.29295	98.81898
90	92.20691	96.21288	97.14683	97.30255	98.87893
100	92.37528	96.22897	97.0867	97.28267	98.78912

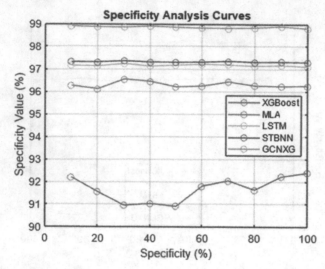

Fig. 5. Specificity values curves comparison among existing and proposed approach

5   Conclusion

The goal of this project is to identify fraudulent behaviours within financial networks using a comprehensive fusion method that combines the skills of graph analytics and machine learning. Our approach demonstrates an improved capacity to identify dishonest behaviours by employing Graph Convolutional Networks (GCNs) to capture intricate structural linkages and integrating the generated node embeddings with Gradient Boosting using XGBoost. Our model is able to recognize intricate patterns that might defy standalone approaches because to the synergistic combination of graph-based insights and iterative boosting processes. There are a number of promising future directions for additional study and development. Extending our methodology to dynamic financial networks, which can support real-time adjustments to node properties and interactions, is a significant direction. Equally potential directions include incorporating unsupervised learning strategies, investigating ensemble methodologies, and assuring model explainability. Furthermore, the approach's applicability and fairness will be further improved by the incorporation of behavioral biometrics, user profiling, and ethical issues. This fusion strategy is a solid solution that successfully integrates the complexity of financial networks with cutting-edge predictive modelling approaches as the landscape of financial fraud continues to change. These improvements will strengthen the integrity of contemporary financial systems while also making a substantial contribution to the fields of financial security and fraud prevention.

References

Smith, A.B., Johnson, C.D., Williams, E.F.: Graph-based anomaly detection for fraud detection in financial networks. J. Financ. Data Sci. **3**(2), 156–168 (2021)

Jin, S., Li, W., Wang, F.: Graph convolutional networks for fraud detection in financial transactions. In: Proceedings of the International Conference on Data Mining (ICDM) (2022)

Chen, J., Li, X., Leung, K.S.: Graph neural networks for fraud detection: a deep learning architecture. IEEE Trans. Big Data **6**(4), 683–692 (2020)

Li, Y., Zhang, Q., Wang, X.: Variational graph autoencoders for anomaly detection in financial networks. Expert Syst. Appl. **190**, 117542 (2023)

Kim, H., Park, J., Lee, D.: Real-time fraud detection in financial networks: an incremental learning approach. ACM Trans. Intell. Syst. Technol. **14**(1), 1–21 (2022)

Hajek, P., Abedin. M.Z., Sivarajah, U.: Fraud detection in mobile payment systems using an XGBoost-based framework. Inf. Syst. Front. 1–19 (2022). https://doi.org/10.1007/s10796-022-10346-6

Zhou, H., Chai, H.F., Qiu, M.L.: Fraud detection within bankcard enrollment on mobile device-based payment using machine learning. Front. Inf. Technol. Electron. Eng. **19**, 1537–1545 (2018). https://doi.org/10.1631/FITEE.1800580

Benchaji, I., Douzi, S., Ouahidi, B.E., Jaafari, J.: Enhanced credit card fraud detection based on attention mechanism and LSTM deep model. J. Big Data **8**, 1–21 (2021). https://doi.org/10.1186/s40537-021-00541-8

Cheng, D., Xiang, S., Shang, C., Zhang, Y., Yang, F., Zhang, L.: Spatio-temporal attention-based neural network for credit card fraud detection. In: Proceedings of the AAAI Conference on Artificial Intelligence, vol. 34, no. 1 (2020). https://doi.org/10.1609/aaai.v34i01.5371

Varalakshmi, A., Santhosh Kumar, S., Shanmugapriya, M.M., Mohanapriya, G., Anand, M.C.J.: Markers location monitoring on images from an infrared camera using optimal fuzzy inference system. Int. J. Fuzzy Syst. **25**, 731–742 (2023). https://doi.org/10.1007/s40815-022-01407-8

Bharatraj, J., Anand, M.C.J.: Power harmonic weighted aggregation operator on single-valued trapezoidal neutrosophic numbers and interval-valued neutrosophic sets. In: Kahraman, C., Otay, İ. (eds.) Fuzzy Multi-criteria Decision-Making Using Neutrosophic Sets. Studies in Fuzziness and Soft Computing, LNCS, vol. 369, pp. 45–62. Springer, Cham (2019). https://doi.org/10.1007/978-3-030-00045-5_3

Anand, M.C.J., Bharatraj, J.: Interval-valued neutrosophic numbers with WASPAS. In: Kahraman, C., Otay, I. (eds.) Fuzzy Multi-criteria Decision-Making Using Neutrosophic Sets. Studies in Fuzziness and Soft Computing, LNCS, vol. 369, pp. 435–453. Springer, Cham (2019). https://doi.org/10.1007/978-3-030-00045-5_17

Anand, M.C.J., Bharatraj, J.: Gaussian qualitative trigonometric functions in a fuzzy circle. Adv. Fuzzy Syst. **2018**, 1–9 (2018). https://doi.org/10.1155/2018/8623465

Anand, M.C.J., Bharatraj, J.: Theory of triangular fuzzy number. In: Proceedings of NCATM – 2017, pp. 80–83 (2017)

Miriam, M.R., Martin, N., Anand, M.C.J.: Inventory model promoting smart production system with zero defects. Int. J. Appl. Comput. Math. **9** (2023). https://doi.org/10.1007/s40819-023-01524-4

Devadoss, A.V., Anand, M.C.J., Felix, A.: A CETD matrix approach to analyze the dimensions of the personality of a person. In: 2014 International Conference on Computational Science and Computational Intelligence. IEEE (2014). https://ieeexplore.ieee.org/abstract/document/6822301

Anand, M.C.J., Moorthy, C.B., Sivamani, S., Indrakumar, S., Kalaiarasi, K., Barhoi, A.: Fuzzy intelligence inventory decision optimization model of sustainability and green technologies for mixed uncertainties of carbon emission. In: 2023 International Conference on Information Management (ICIM). IEEE (2023). https://ieeexplore.ieee.org/document/10145085

Anand, M.C.J., Martin, N., Clementking, A., Rani, S., Priyadharshini, S., Siva, S.: Decision making on optimal selection of advertising agencies using machine learning. In: 2023 International Conference on Information Management (ICIM). IEEE (2023). https://ieeexplore.ieee.org/document/10145172

Raj, P.J., Prabhu, V.V., Krishnakumar, V., Anand. M.C.J.: Solar powered charging of fuzzy logic controller (FLC) strategy with battery management system (BMS) method used for electric vehicle (EV). Int. J. Fuzzy Syst. (2023). https://doi.org/10.1007/s40815-023-01537-7

Prabha, S.K., et al.: Sorting out interval valued neutrosophic fuzzy shortest cycle route problem by reduced matrix method. Int. J. Neutrosophic Sci. 23(2), 91–103 (2024). https://doi.org/10.54216/IJNS.230208

Kungumaraj, E., et al.: Neutrosophic topological vector spaces and its properties. Int. J. Neutrosophic Sci. 23(2), 63–76 (2024). https://doi.org/10.54216/IJNS.230206

Rajesh, K., et al.: A study on interval valued temporal neutrosophic fuzzy sets. Int. J. Neutrosophic Sci. 23(1), 341–349 (2024). https://doi.org/10.54216/IJNS.230129

Manshath, A., et al.: Neutrosophic integrals by reduction formula and partial fraction methods for indefinite integrals. Int. J. Neutrosophic Sci. 23(1), 08–16 (2024). ISSN: 2690-6805. https://doi.org/10.54216/IJNS.230101

Sudha, S., Martin, N., Anand, M.C.J., Palanimani, P.G., Thirunamakkani, T., Ranjitha, B.: MACBETH-MAIRCA plithogenic decision-making on feasible strategies of extended producer's responsibility towards environmental sustainability. Int. J. Neutrosophic Sci. 22(2), 114–130 (2023). ISSN: 2690-6805. https://doi.org/10.54216/IJNS.220210

Power Distribution System Power Quality Enhancement with Custom Power Devices Utilizing Machine Learning Techniques

N. Raveendra[1]([⊠]), A. Jayalaxmi[2], and V. Madhusudhan[3]

[1] St. Peters Engineering College, Dhullapally, Secunderabad, Telangana, India
nraveendra@stpetershyd.com
[2] JNTUH College of Engineering, Kukatpally, Hyderabad, Telangana, India
[3] VNRVJIET, Bachupally, Telangana, India

Abstract. An issue with power quality is one that arises from an abrupt increase in an abnormal voltage, current, or frequency. Poor power quality or non-linear loads can lead to a distribution system's voltage sag, swell, interruptions, harmonics, and transients, among other issues. Various compensating devices are utilized nowadays to enhance power quality. To provide quick, adaptable, and effective solutions for different power disturbances which include devices like DVR Voltage Re-storer and DSTATCOM are taking into consideration advancements in power electronic technologies like converter with various magnitudes. These tools rectify magnitude, current, source disturbances brought on by various defects and loads. In order to test it and lower total harmonic distortion, they are connected to the main distribution network using the IEEE 14 Bus standard. To improve power quality in the utility, the use of sophisticated instruments for power quality analysis is becoming more and more important every day. In order to reduce harmonics in bus voltages and bus currents, DSTATCOM and DVR employ optimal PI-based neural network techniques and linear regression techniques, which are analyzed and compared in this study.

Keywords: Power Quality · DSTATCOM · DVR · custom power devices · Linear load · non-linear load · SPWM · Total harmonic distortion (THD) · NN Tool · Linear Regression

1 Introduction

Numerous quantities, including voltage, current, and others, are used to measure power quality. If the power distribution system gives its consumers a constant supply of energy and a voltage that is lossless, then it is a good and dependable system. However, in actual use, distribution systems feature a variety of nonlinear (diode bridge and induction motor) and linear (balanced and unbalanced) loads, substantial effect on quality of consumed. The term "power quality disturbance" departs magnitudes from their ideal values. Transmission or distribution faults can induce voltage sags or swells across the

© The Author(s), under exclusive license to Springer Nature Switzerland AG 2024
S. L. Gundebommu et al. (Eds.): REGS 2023, CCIS 2081, pp. 33–50, 2024.
https://doi.org/10.1007/978-3-031-58607-1_3

entire system, or just a significant portion of it. The system may have a considerable voltage decrease under heavy load circumstances, whereas the voltage increases noticeably under light load conditions. Numerous disturbances, including disturbances of magnitudes, can cause Power Quality difficulties [1]. The majority of problems like voltage sags/surges can happen. Significant issues with the system are these sags and swells. Different compensatory technologies, such as DSTATCOM and DVR, are utilized to address these issues [2]. Major power quality issues like sag and swell, as well as minor issues like flickers, harmonics, transients, etc., are lessened with the use of these devices [3].

These two devices are used to suggest various control schemes for load voltage regulation. Voltage control and reactive power correction are carried out earlier in DSTATCOM. Compensation open-loop, closed-loop load voltage regulation are carried out earlier in DVR. The two devices are thought to operate in the best, most accurate, and fastest closed-loop voltage control mode when it comes to handling abrupt changes in both the voltages. The voltage-control in feedback mode applied for both compensators. In this work, several combinations of bespoke switching devices, like DSTATCOM and DVR, in conjunction with various types of loads, are used to effectively implement various power performance concerns, such as different. Disturbances.

2　Custom Power Devices

Power electronic-based switches which enhance control electric transmission networks with different scenarios. The standard of the supply should be met. Otherwise, decreased utility output may occasionally result in an entire industry closure, which would create a large financial loss for the sector. It has been noted that the majority of circumstances in industries that can impede the process occur within the load area. Result in transients, which may adapt with power supply's stability, quality, and dependability. The few aberrant electrical circumstances listed below can prevent the Transmission and Distribution (T&D) system from operating in a healthy manner: Voltage Sag (Dip):

a. Extremely brief disruption or stoppage of time
b. Swell and voltage spikes: Prolonged disruption or disturbance.
c. Unbalanced and fluctuating voltage: Noise and harmonic distortion.
d. Custom power device controllers fall into one of three categories based on the kind of network connection they have:
e. Controllers with shunt connections: linked in parallel adjustment.
f. Controllers connected: linked to the power system in series (series compensation)
g. A power system with hybrid connected.
h. Hybrid controllers combined with power system connections.

2.1　Distribution Static Compensator (DSTATCOM)

The Distribution Static Compensator (DSTATCOM) is a significant shunt compensator among the different distribution controllers. It possesses the ability to address power quality issues encountered by distribution systems. [4]. A Static VAR Compensator (SVC) has been essentially superseded by DSTATCOM since it requires a significant reaction time. It pertains to passive filter banks and is limited to reactive power compensation under steady state conditions. A FACTS controller like DSTATCOM, that is based on a Voltage Source Inverter (VSI) and shares ideas with a STATCOM that is utilized at the transmission level. Furthermore, because SVCs only partially reduce instantaneous flicker level, DSTATCOMs have taken the role of SVCs, which were Primarily employed in arc welding facilities to mitigate voltage fluctuations. In essence, a DSTATCOM is a FACTS controller based on a Voltage Source Converter (VSC) that shares many characteristics with a STATCOM utilized at the transmission level.

Whereas a FACTS controller is used for optimization which includes voltage magnitudes. When it comes to eliminating distortion and imbalance in magnitudes. DSTATCOMs perform similarly to shunt active filters. The technique determines how well the DSTATCOM performs. Different theories and techniques, such as phase shift control, instantaneous PQ theory, and synchronous frame theory, provide the basis for different control algorithms for the DSTATCOM block [5].

Every one of the above algorithms has advantages and disadvantages of its own. Five control mechanisms have been used in this paper for controlling power. The approach is employed to improve the performance of power transmission systems. Alternative control strategies for compensating the nonlinear power electronic load and imbalanced linear load are hysteresis control methods, synchronous frame instantaneous PQ theory. A typical distribution power transformer connects the voltage source inverter, known as DSTATCOM, distribution network. It is used to adjust bus voltage sags. The DSTATCOM provides leading or lagging compensatory current while continually monitoring the line waveform. Figure 1 displays the DSTATCOM single line diagram. A distribution transformer, an L-C filter, one or more converter modules, a DC capacitor, and a PWM control are the components involved in this setup. Method makes up DSTATCOM [6]. Through the use of a tiny A voltage-source inverter consisting of a tied capacitor and reactor (filter L-C) synchronizes and connects DC to AC.

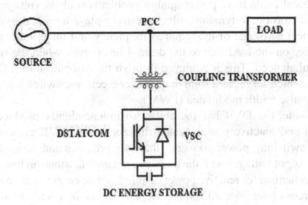

Fig. 1. Schematic diagram of DSTATCOM.

FACTS devices are as below:

- $V_i > V_M$.
- $V_i < V_M$.
- $V_i = V_M$, which is sometimes referred to as a balanced situation.

Volts in the inverter, expressed as V_i
V_M is the voltage in volts
Source voltage in volts is Vs.

Every control algorithm determines the compensator's compensated current in order to either provide or absorb reactive power. This yields the compensated current.

$$I_s = I_L = I_c \tag{1}$$

where, I_c = Current Compensated (A)
I_L = Current Load (A)
I_s = Current Source (A)
VSC is utilized in conjunction with the DC storage capacitor in this arrangement. One leg uses two IGBT switches, three legs in parallel. Compensatory are filtered through the interface between the filter resistance Rf and the inductance Lf. The converter's switching frequency is set by the value of its inductance, Lf.

2.2 Dynamic Voltage Restorer (DVR)

A static variable device with numerous uses in distribution and transmission networks is the dynamic voltage restorer (DVR) [7]. It is a series compensation device that uses power electronic controllers that employ Voltage Source Converters (VSC) In order to protect vulnerable electrical loads from power quality problems such as voltage sags, swells, imbalance, and distortion, a dynamic voltage restorer plays a crucial role. Its primary function is to inject a voltage of the required frequency and magnitude, allowing it to restore the voltage on the load side to the desired level, even when the source voltage is unstable or unbalanced. This is achieved through the utilization of a Gate Turn Off (GTO) thyristor, which serves as a solid-state power electronic switch within an inverter structure that is pulse width modulated (PWM).

At the load side, the DVR has the ability to independently produce and absorb controllable real and reactive power. Stated differently, the DVR consists of a solid-state DC to AC switching power converter that synchronizes and series-feeds a set of three-phase AC output voltages with the transmission and distribution line voltages. The commutation mechanism for reactive power demand and an energy source for real power need are the sources of the injected voltage. Depending on the DVR's manufacturer and design, the energy source may change. Applied energy sources include DC capacitors, batteries, and energy obtained from the line via a rectifier.

The warped, non-linear sinusoidal signal's FFT spectrum. At the signal's fundamental frequency, f1, the biggest peak is visible. At integral multiples of the fundamental frequency (f2 = 2·f1, f3 = 3·f1, f4 = 4·f1, etc.), additional peaks can be seen. The peaks, which are referred to as harmonics, show how a signal behaves nonlinearly. Because of the increased non-linear loads, the harmonics' magnitude rises. The so-called THD factor can be computed using the obtained FFT data. A measurement of the harmonic distortion of a signal is called the total harmonic distortion (Fig. 2 and Table 1).

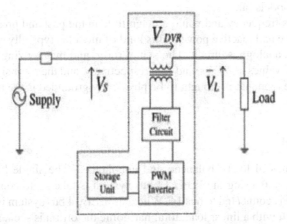

Fig. 2. Schematic diagram of DVR

Table 1. Total Harmonic Distortion (THD) with custom power devices

S. No	FACTS devices (Without Load)	Voltages THD (%)			Avg Voltage THD (%)
		Phase a	Phase b	Phase c	
1	DVR	1.88	1.59	1.61	1.693
2	DSTATCOM	4.84	3.38	4.58	4.26

Analysis: The tabular findings show that FFT analysis is used to determine the overall harmonic distortion for three phase voltages for DSTATCOM and DVR.

3 Types of Loads

Power system studies are significantly impacted by load modeling. A review of load modeling and identification methods is given in this study. Static and dynamic models are the two main categories into which load models fall, and measurement-based and component-based approaches are the two ways in which model parameters are determined [8].

Demand-side management, smart metering, and the integration of renewable energy sources have drawn increased attention to load modeling in recent years. The commonly employed load models, however, are out of date and unable to capture growing loads. In order to satisfy the growing demand from both business and academia, it is necessary to thoroughly analyze the load modeling methods that are now in use and recommend future research areas. A thorough examination of the state of academic research, industrial practices, and load modeling's current problems and emerging trends was conducted. The mathematical depiction of the relationship between voltage and power in a load bus is known as load modeling. Static and dynamic models are the two primary categories into which load models fall.

The load bus's frequency and voltage magnitude in the past and present are used to represent the active and reactive power. This kind of model is typically created using an induction motor's analogous circuit. The static, rotor, and magnetizing resistances are represented by the values of Rs, Xr, and Xm, respectively, and the rotor slip is denoted by the letter "s." One type that is thought to be physically grounded is the induction motor model.

3.1 Linear Load

There are two kinds of loads: balanced and unbalanced. The single line diagram for power quality analysis using the IEEE 14 bus system is taken into consideration. The load in the system is connected to bus 14. When an IEEE 14 bus system is coupled to the FFT analysis result with a linear load, total harmonic distortion is seen. It demonstrates how the system's distortion is extremely low with a linear load since current varies in direct proportion to voltage. A three-phase series RLC load is regarded as a linear load in this instance, and the current in such a load varies linearly with the voltage. Here, the voltage at bus 14 is used to calculate the THD. This outcome demonstrates that, for a linear load, the overall harmonic aberrations in the line voltage are extremely small, or insignificant.

3.2 Non-linear Load

One of the main issues with power quality is harmonics. The waveforms of voltage and current are distorted by harmonics, which has a negative impact on electrical equipment. The following are some instances of nonlinear loads:

- Diode Bridge converter
- Modifiable drive mechanisms, such as induction motors

Bus 14 is where the IEEE 14 bus system with nonlinear load is connected. When studying power quality, linear loads such as an induction motor and a diode bridge rectifier are taken into account in steady state. When the load is connected to the IEEE 14 bus system, harmonics with a nonlinear load are studied in this case. After connecting the diode bridge rectifier to bus 14 as a nonlinear load, power quality is examined using FFT analysis. The result with the diode bridge rectifier is shown in FFT analysis. It is common knowledge that every industry requires a different amount of power from generating firms or other utilities in order to meet its varying load requirements at the terminal. An induction motor is a major source of power quality issues in many sectors. Thus, the induction motor-related power quality issue is investigated, and using FFT analysis, the Total Harmonic Distortion (THD) is noted. With linear (balanced and

Table 2. Total Harmonic distortion (THD) with FACTS devices, Loads

S. No	FACTS Devices	Type of Load	Voltage THD (%)
1	DSTATCOM	Linear Load - Balanced	0.28
		Linear Load - Unbalanced	0.40
		Non-Linear Load – Diode bridge	1.20
		Non- Linear Load – Induction motor	3.6
2	DVR	Linear Load - Balanced	1.94
		Linear Load - Unbalanced	9.3
		Non-Linear Load – Diode bridge	13.69
		Non- Linear Load – Induction motor	14.75

unbalanced) and non-linear (diode bridge and induction motor) loads, DSTATCOM and DVR are used (Table 2).

Analysis: The tabular findings show that FFT analysis is used to determine the total harmonic distortion for voltages for DSTATCOM and DVR with varying loads.

4   Multi-level Inverter with SPWM

The best application of multilevel inverter technology is now seen in high-power medium voltage drives and transmissions. A comparison investigation is done with the performance of 2-level and 3-level inverters linked to induction motor for power quality objectives. By removing the loud harmonics, these schemes can have higher power quality. A passive filter is thoughtfully designed to lower the overall harmonic distortion of the voltage waveform that results [9].

Active filtering has become a mature method for improving power quality in recent years thanks to a new generation of voltage source converters known as multilayer inverters, which are capable of creating a desired voltage from many levels of DC voltage as inputs [10]. A multi-level converter has less harmonic distortion and switching stress than a two-level converter. While producing the optimal output voltage, the switching approach used to regulate the inverter effectively contributes to harmonic removal. The optimal control scheme should be chosen in order to minimize total harmonic distortion, after extensive research on topology and control approaches.

[11] In order to adjust for voltage sag and voltage swell in power distribution networks, this research suggests basic 2-level and 3-level multilevel inverters using DSTATCOM and DVR. A multilevel topology with separated DC energy storage and fewer switches is used to construct the suggested DSTATCOM (Fig. 3).

To fix the voltage sag, swell, and interruption, a DVR injects a voltage in series with the system voltage and a DSTATCOM implants a current into the system. The generation of a firing pulse for a multi-level inverter is explained using the sinusoidal PWM technique. MATLAB/Simulink is utilized to simulate the suggested methodology.

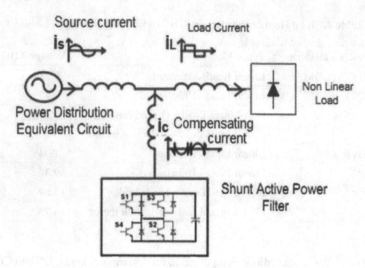

Fig. 3. Schematic diagram of Shunt active power filter

The width of each pulse is varied in proportion to the amplitude of a sine wave assessed at the center of the same pulse, as opposed to all pulses having the same width as in the case of multiple pulse width modulation.

- Line to ground voltage (supply voltage): –415 V;
- supply frequency: –50 Hz; VSI switching frequency: 10 kHz;
- coupling inductor: 3 mH;
- DC side capacitor: 1000 μF;
- Vdc, ref. = 650 V; active load power: 1 KW
- 6 pulse diode rectifier for non-linear load;
- 1 KVA for reactive load power

Two control strategies—a stationary reference frame and a revolving frame—are used to minimize total harmonic distortion (THD) in two-level and three-level MLI with SPWM (Figs. 4 and 5, Tables 3, 4).

Analysis: The tabular column shows that, in comparison to a three-level MLI, the THD of a DSTATCOM device with a two-level MLI is lower.

The DSTATCOM device is utilized in the IEEE standard 14 Bus system to mitigate voltage sag at Bus 14, according to the simulation models and findings. By using neural network technology to alter hidden layers, voltage sag is lessened. A classic backpropagation learning algorithm, which minimizes the Mean Squared Error (MSE) of the training data, is frequently used to train these supervised neural networks.

5 Machine Learning Techniques

Electric power utilities have established intensive data gathering initiatives to evaluate the Power Quality (PQ) problems in their systems [11]. Handling the vast amounts of monitored data will be made easier with the use of machine learning for the classification

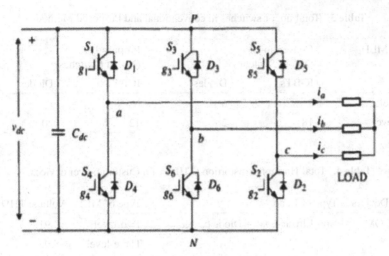

Fig. 4. Two-Level Inverter

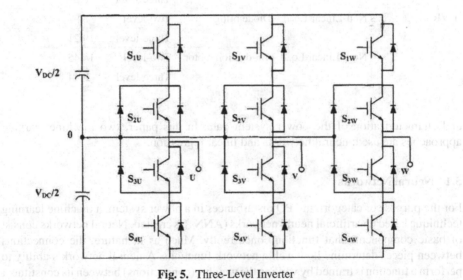

Fig. 5. Three-Level Inverter

Table 3. Total no. of switches in conventional and Proposed Method

Type of MLI	Conventional (No. of Switches)		Proposed (No. of Switches)	
	IGBTs	Diodes	IGBTs	Diodes
Two-level	9	6	6	-
Three-level	18	26	12	20

Table 4. Total Harmonic distortion (THD) with Custom power devices

FACTS Devices	Type of Load	Type of MLI	Voltage THD (%)
DSTATCOM	Non-Linear Load – Diode bridge	Two-level	2.91
		Three-level	8.92
	Non- Linear Load – In- duction motor	Two-level	4.25
		Three-level	6.23
DVR	Non-Linear Load – Diode bridge	Two-level	8.94
		Three-level	19.21
	Non- Linear Load – In- duction motor	Two-level	14.45
		Three-level	9.44

and characterization of the power system data. In this paper, two machine learning approaches are used: neural networks and linear regression.

5.1 Neural Networks

For the purpose of categorizing PQ disturbances in a power system, a machine learning technique based on artificial neural networks (ANNs) is created. Neural networks consist of basic components that function concurrently. Much as in nature, the connections between pieces determine how well a network functions. A neural network's ability to perform a function is trained by varying the weights (connections) between its constituent pieces. Neural networks are frequently trained, or altered, to ensure that a given input results in a desired output. After that, the network is changed such that the output of the network and the target match, based on a comparison of the two. In supervised learning, a large number of these input/tar-get combinations are often utilized to train a network (Fig. 6).

When training a network in batches, weight and bias adjustments are made using all of the input vectors in the batch. Gradient training presents each individual input vector to a network and adjusts its weights and biases as appropriate. Sometimes, "on-line" or "adaptive" training is used to describe incremental training. Many domains of application, such as pattern recognition, identification, classification, speech, vision, and control systems, have trained neural networks to carry very sophisticated tasks. Neural

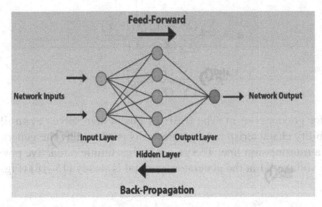

Fig. 6. Block diagram of Neural Network

networks are now able to be trained to handle issues that humans and conventional computers find challenging. While supervised training approaches are frequently employed, direct design methods or unsupervised training approaches can provide alternative networks. For example, unsupervised networks can be used to find data clusters. Some linear network types, such as the backpropagation technique, are created directly.

5.2 Linear Regression

Neural networks are now able to be trained to handle issues that humans and conventional computers find challenging. While supervised training approaches are frequently employed, direct design methods or unsupervised training approaches can provide alternative networks. For example, unsupervised networks can be used to find data clusters. Some linear network types, such as the backpropagation technique, are created directly.:

$$Y_i = b + aX_i + \epsilon_i, \quad i = 1, \dots, n, \qquad (2)$$

where a, b represents the intercept and slope of the line, respectively, and Y_i, X_i are the values of the response and predictor variables in the trial; the unknown parameters are typically assumed to be δ_{id} (error) from N $(0, \sigma^2)$, particularly for inference purposes. Subsequently, estimations of the parameters of the basic linear model has to be acquired appropriately, employing a technique such as the ordinary least squares method, which depends on reducing the sum of squared errors. The ordinary least squares estimate of a and b for the basic linear regression model are as follows:

Where Yi, Xi are the values of the response and predictor variables in the trial, respectively; the unknown parameters: a is called the intercept, and b is the slope of the line; ϵ_i are usually assumed to be δid (error) from N $(0, \sigma^2)$ specially for inference purposes. Then estimates of simple linear model parameters should be obtained accordingly, using some method the ordinary least squares method, which relies on minimizing the sum of square of error. For the simple linear regression model, the ordinary least squares estimations of a and b are:

$$\hat{b} = \bar{Y} - \hat{a}\bar{X} \qquad (3)$$

$$\hat{a} = \frac{\sum_i (X_i - \overline{X})(Y_i - \overline{Y})}{\sum_i (X_i - \overline{X})^2}. \tag{4}$$

$$R^2 = \frac{\sum (\hat{Y}_i - \overline{Y})^2}{\sum (Y_i - \overline{Y})^2}. \tag{5}$$

R^2 stands for goodness of fit. The IEEE 14-bus system serves as an illustration of the voltage stability characteristic. Active power is produced by the generator and sent to the load via a transmission line. The generator has limitless reactive power capacity. As a result, the voltage V1 at the generator terminal is steady [15–18] (Fig. 7).

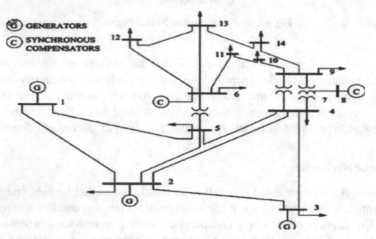

Fig. 7. IEEE Standard 14 Bus system

The equations for voltages at buses (14) are given by,

$$V_2 = \frac{\sqrt{((V_1^2 - 2QX) \pm \sqrt{(V_1^4 - 4QXV_1^2 - 4PX^2))}}}{2} \tag{6}$$

$$P_i = \sum V_i V_j Y_{ij} \cos (\delta_i - \delta_{ij} Y_{ij})$$
$$Q_i = \sum V_i V_j Y_{ij} \sin (\delta_i - \delta_{ij} Y_{ij}) \tag{7}$$

For $i, j = 1$ to n, where Pi and Qi are active and reactive powers injected at load (Figs. 8 and 9).

Analysis: Active and reactive power variation w.r.t time for the bus 14 where the volt age deviation observed is less than 1 p.u. and Training of NN network is observed in Fig. 10.

Analysis: Figure 11 compares the regression coefficient from NN network and linear regression (Figs. 13 and 14).

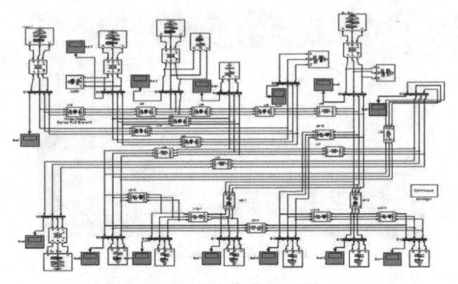

Fig. 8. Simulation Model for IEEE 14-BUS system

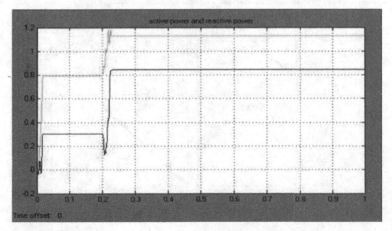

Fig. 9. Active power and Reactive power Versus time (secs)

Analysis: Training of NN network is observed in Fig. 12 for obtaining best performance in terms of Mean Square Error (MSE) (Table 5).

Analysis: Voltage sag at Bus 14 in pu is mitigated by back propagation technique with variation of Hidden Layers. The MSE and R are the parameters analyzed for improving accuracy of the system (Table 6).

Analysis: The function of linear regression analysis is to compare the actual outputs of the neural network with the corresponding target outputs. The closer the value of R to 1, more accurate the prediction.

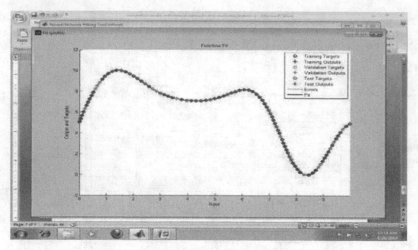

Fig. 10. Results after training

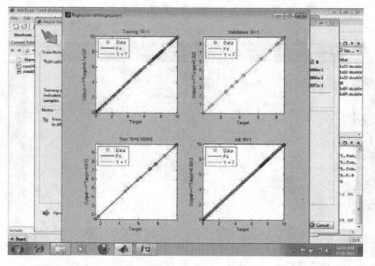

Fig. 11. MSE value comparison with NN, Linear Regression

Analysis: From Table 7, it is observed that Neural network is comparatively better for harmonic reduction in voltage than linear regression for both Non-linear loads like diode bridge, Induction motor. When the results of two FACTS devices are compared, DSTATCOM has comparatively less % Total Harmonic Distortion than compared to DVR.

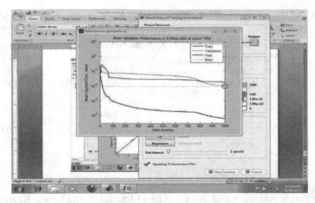

Fig. 12. Comparing results of performance with NN and Linear regression

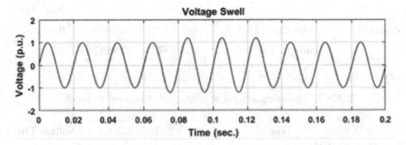

Fig. 13. Voltage sag (Volts) Versus Time (secs)

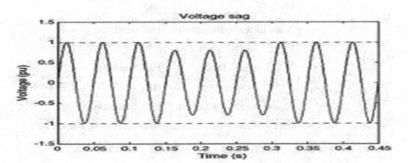

Fig. 14. Voltage Swell (Volts) Versus Time (secs)

Table 5. Voltage Sag obtained at Bus 14 with Back Propagation Algorithm.

Hidden layers	MSE (Mean square error)	Regression coefficient (Accuracy)	Voltage at Bus 14 (P.U)
1	$6.40e^{-6}$	0.99544	0.82
2	$3.074e^{-5}$	0.99688	0.85
5	$1.13e^{-4}$	0.99999	0.76
8	$5.45e^{-5}$	0.98088	0.78
10	$1.00e^{-5}$	0.9969	1.07
15	$4.18e^{-3}$	0.513	1.02

Table 6. Comparison of RMSE, R and Voltage at Bus 14 with NN and Linear Regression methods

Methods		RMSE	R	Voltage at Bus 14 (P.U)
Linear Regression	Basic Curve Fitting	0.0403	0.7706	0.89
	Full quadratic	0.01952	0.9502	0.92

Table 7. Comparison table with Machine Learning methods

Custom power Devices (With Two- level, SPWM)	Type of Load	Type of MLI	Voltage THD (%)
DSTATCOM	Non-Linear – Diode bridge	Full quadratic	3.45
		NN	1.53
	Non- Linear – Induction motor	Full quadratic	4.4
		NN	1.28
DVR	Non-Linear – Diode bridge	Full quadratic	4.36
		NN	1.99
	Non- Linear – Induction motor	Full quadratic	3.98
		NN	1.48

6   Conclusion

This study-built algorithms that were preset in MATLAB/Simulink using an IEEE standard 14 bus distribution system with a variety of loads, specific power devices, and multi-level inverters. For both linear and non-linear loads, a two-level inverter has two advantages over a three-level inverter: low harmonic distortion and fewer switches. The most effective power quality (PQ) compensator for reducing PQ issues in a distribution system is the distribution static compensator (DSTATCOM). The performance of the PQ compensator under nonideal source conditions and fluctuating load is determined by the control strategy. This study uses machine learning techniques to reduce voltage sag at Bus 14 of the IEEE standard test system. The voltage THD in the neural network is around 45% lower than the voltage THD produced by the full quadratic linear regression technique, according to the simulation findings.

References

1. Stones, J., Collinson, A.: Power quality. Power Eng. J. **15**, 58–64 (2001)
2. Haque, M.H.: Compensation of distribution system voltage sag by DVR and D-STATCOM. In: 2001 IEEE Porto Power Tech Proceedings (Cat. No. 01EX502) (Vol. 1, pp. 5-pp). IEEE (2001)
3. Ghosh, A., Ledwich, G.: Power Quality Enhancement Using Custom Power Devices. Springer, New York (2012). https://doi.org/10.1007/978-1-4615-1153-3
4. Mishra, M.K., Ghosh, A., Joshi, A.: Operation of a DSTATCOM in voltage control mode. IEEE Trans. Power Deliv. **18**(1), 258–264 (2003)
5. Ismail, N., Abdullah, W.N.W.: Enhancement of power quality in distribution system using D-STATCOM. In: 2010 4th International Power Engineering and Optimization Conference (PEOCO), pp. 418–423. IEEE (2010)
6. Mindykowski, J., Tarasiuk, T., Rupnik, P.: Problems of passive filters application in system with varying frequency. In: 2007 9th International Conference on Electrical Power Quality and Utilisation, pp. 1–4. IEEE (2007)
7. Zhi, L.X., Yin, Z.D., Ding, H., Han, J.B., Hu, F.X.: Study on energy-saving strategies for dynamic voltage restorer. In: 2006 International Conference on Power System Technology, pp. 1–5. IEEE (2006)
8. Barakati, S.M., Sadigh, A.K., Mokhtarpour, E.: Voltage sag and swell compensation with DVR based on asymmetrical cascade multicell converter. In: North American Power Symposium (NAPS), pp. 1–7 (2011)
9. Mahela, O.P., Khan, B., Alhelou, H.H., Tanwar, S.: Assessment of power quality in the utility grid integrated with wind energy generation. IET Power Electron. **13**(13), 2917–2925 (2020)
10. Zhong, T., Zhang, S., Cai, G., Huang, N.: Power-quality disturbance recognition based on time-frequency analysis and decision tree. IET Gener. Transm. Distrib. **12**(18), 4153–4162 (2018)
11. Zhong, T., Zhang, S., Cai, G., Li, Y., Yang, B., Chen, Y.: Power quality disturbance recognition based on multiresolution s-transform and decision tree. IEEE Access **7**, 88380–88392 (2019)
12. Ibrahim, S., Daut, I., Irwan, Y.M., Irwanto, M., Gomesh, N., Farhana, Z.: Linear regression model in estimating solar radiation in Perlis. Energy Procedia **18**, 1402–1412 (2012)
13. Lin, L., Wang, D., Zhao, S., Chen, L., Huang, N.: Power quality disturbance feature selection and pattern recognition based on image enhancement techniques. IEEE Access **7**, 67889–67904 (2019)
14. Biswas, P.P., Arora, P., Mallipeddi, R., Suganthan, P.N., Panigrahi, B.K.: Optimal placement and sizing of FACTS devices for optimal power flow in a wind power integrated electrical network. Neural Comput. Appl. 1–22 (2020)
15. Rajeswaran, N., Swarupa, M.L., Maddula, R., Alhelou, H.H., Kesava Vamsi Krishna, V.: A study on cyber-physical system architecture for smart grids and its cyber vulnerability. In: Haes Alhelou, H., Hatziargyriou, N., Dong, Z.Y. (eds.) Power Systems Cybersecurity, pp. 413–427. Power Systems. Springer, Cham (2023). https://doi.org/10.1007/978-3-031-20360-2_17
16. Vostriakova, V., Swarupa, M.L., Rubanenko, O., Gundebommu, S.L.: Blockchain and climate smart agriculture technologies in agri-food security system. In: Kumar, A., Fister Jr., I., Gupta, P.K., Debayle, J., Zhang, Z.J., Usman, M. (eds.) Artificial Intelligence and Data Science. ICAIDS 2021. CCIS, vol. 1673, pp. 490–504. Springer, Cham (2022). https://doi.org/10.1007/978-3-031-21385-4_40

17. Swarupa, M.L., Divya, G., Lavanya, V.C.S.N.: Multi-agent system for energy management of renewable energy in domestic cooking. In: Mahajan, V., Chowdhury, A., Padhy, N.P., Lezama, F. (eds.) Sustainable Technology and Advanced Computing in Electrical Engineering. LNEE, vol. 939, pp. 453–468. Springer, Singapore (2022). https://doi.org/10.1007/978-981-19-4364-5_33
18. Swarupa, L., Lakshmi, S., Reddy, K.: Virtual Power Plant Solution for Future Smart Energy Communities, 1st ed. CRC Press, Boca Raton (2022). https://doi.org/10.1201/9781003257202

Diabetes Prediction Using Logistic Regression

Zarinabegam Mundargi[1] , Mayur Dabade[2(✉)], Yash Chindhe[2], Savani Bondre[2], and Anannya Chaudhary[2]

[1] Vishwakarma Institute of Technology, Pune, India
zarinabegam.mundargi@vit.edu
[2] Department of Artificial Intelligence and Data Science, Vishwakarma Institute of Technology, Pune, India
{mayur.dabade21,yash.chindhe21,savani.bondre211,
ashish.anannya21}@vit.edu

Abstract. Diabetes mellitus, characterized as a chronic metabolic condition, presents a notable global health concern. Timely detection and intervention play a crucial role in the effective management and enhancement of patient outcomes. This research paper explores the application of logistic regression as a predictive tool for diabetes diagnosis. Leveraging a substantial dataset containing clinical and patient-related variables, our study demonstrates the feasibility and efficacy of logistic regression pinpoint individuals susceptible to developing diabetes. By analyzing relevant features, and calculating the sigmoid function, cost function, and gradient descent from scratch and employing an optimal threshold, the logistic regression model exhibits commendable accuracy, sensitivity, and specificity. These findings highlight its potential as an early diagnostic tool. Such predictive models hold promise for healthcare practitioners, enabling them to proactively identify high-risk individuals and initiate preventive measures. As a cost-effective and accessible method, logistic regression aids in the early diagnosis and management of diabetes, ultimately leading to enhanced healthcare strategies and patient care.

Keywords: Diabetes · Logistic Regression · Gradient descent · Learning rate

1 Introduction

Diabetes is a prevalent chronic ailment that predominantly affects the elderly population worldwide. According to the International Diabetes Federation, the year 2017 witnessed 451 million people grappling with diabetes across the globe, and this number is expected to surge to a staggering 693 million individuals over the next 26 years [1]. Diabetes presents as a persistent condition characterized by fluctuating blood glucose levels. It results from pancreatic dysfunction, leading to either inadequate insulin production (Type 1 diabetes) or reduced cellular responsiveness to insulin (Type 2 diabetes) [2, 3].

The precise cause of diabetes remains incompletely understood, although it is widely recognized that genetic and lifestyle elements are considered as significant factors in its onset.

© The Author(s), under exclusive license to Springer Nature Switzerland AG 2024
S. L. Gundebommu et al. (Eds.): REGS 2023, CCIS 2081, pp. 51–61, 2024.
https://doi.org/10.1007/978-3-031-58607-1_4

While diabetes is incurable, it is manageable through various treatment modalities and medications. Individuals with diabetes are at a heightened risk of developing secondary health complications, such as nerve-related problems and heart diseases. Thus, early detection and intervention are critical in averting complications and mitigating severe health risks.

In recent times, bioinformatics researchers have collectively worked towards addressing the challenge posed by diabetes. They have sought to develop systems and tools for the diabetes prediction, employing a range of machine learning algorithms, including classification and association algorithms, have been extensively explored in the domain of diabetes prediction. Notably, Logistic Regression, Support Vector Machine (SVM), and Decision Trees have emerged as highly prominent choices for predicting diabetes [4].

This study aims to contribute to ongoing efforts by delving into the application of Logistic Regression, a powerful machine learning technique, for diabetes prediction. By developing a predictive model based on clinical data, we aim to facilitate early detection of individuals at risk, ultimately enhancing the prospects for proactive healthcare and targeted interventions in the battle against diabetes.

2 Literature Review

In the realm of diabetes prediction, Numerous investigations have delved into the utilization of machine learning techniques and statistical models to predict diabetes risk, with a focus on enhancing early detection and prevention. Prior research has delved into the application of algorithms like support vector machines, decision trees and logistic regression for diabetes prediction. Moreover, some studies have incorporated advanced techniques, including deep learning and ensemble methods, to improve predictive accuracy. Additionally, the integration of diverse data sources, such as electronic health records and wearable device data, has expanded the scope of diabetes prediction research. This section will provide an overview of these related works, highlighting their methodologies and findings in the context of diabetes risk assessment.

Souad & Aburahmah [5] addresses diabetes prediction using a wide range of Machine Learning and Deep Learning techniques. While existing studies have achieved high accuracy, this paper investigates rarely used ML classifiers, attaining accuracies in the range of 68% to 74%. The survey offers a thorough examination of deep learning and machine learning methodologies in the diabetes prediction over the past six years. The findings suggest the potential for enhancing prediction by combining these classifiers, highlighting avenues for further research and model development in this critical healthcare domain.

Tejas & Pramila [6] confronts the urgent global challenge posed by diabetes, a pervasive chronic ailment impacting millions worldwide. Leveraging the capabilities of data science and machine learning, the investigation strives to elevate the precision of early diabetes prediction. This endeavor hinges on the deployment of three supervised machine learning approaches: SVM, Logistic Regression, and ANN. Given the intricate interconnections within diabetes and its multifaceted impact on various bodily organs, this study endeavors to introduce an innovative healthcare solution for the earlier detection of this disease.

KM Jyoti Rani [7] delves into the pressing issue of diabetes, a widespread chronic disease with serious health implications. Paper highlights the alarming global prevalence and the potential for this to double by 2035. The paper underscores how high blood glucose levels are at the core of diabetes, leading to various symptoms and severe complications, including blindness, kidney failure, and heart disease.

The study focuses on machine learning, an emerging field in data science, to develop a predictive system for early diabetes detection. Leveraging algorithms such as Logistic Regression, Random Forest, Support Vector Machine, K-nearest neighbour, and Decision Tree, the research aims to enhance prediction accuracy. These methods are rigorously evaluated to determine their effectiveness in diabetes prediction.

[8] This research addresses diabetes, a major global health concern associated with severe complications. It employs data mining techniques, including Support Vector Machine, Back Propagation, and Naive Bayes to predict diabetes. Using an 8-parameter input layer, a 6-neuron hidden layer, and an output layer, the study achieves 83.11% accuracy, 86.53% sensitivity, and 76% specificity with the Back Propagation algorithm. This represents an improvement over previous work. Comparative analysis with other algorithms validates these findings.

Salliah Shafi Bhat et al. [9] focused on diabetes prediction, a significant global health issue linked to severe complications. Leveraging Machine Learning Algorithms (MLA), the study aims to enhance early disease identification. Real clinical data from Bandipora, India, collected between April 2021 and February 2022, was analyzed. Six MLAs were employed, with Random Forest (RF) demonstrating the highest accuracy at 98%, followed by SVM (92%), MLP (90.99%), DT (96%), GBC (97%), and LR (69%). This research contributes to the effective identification of diabetes prevalence and prediction.

Nahzat and Yağanoğlu [10] explored the application of machine learning classification algorithms for diabetes prediction in their paper titled 'Diabetes Prediction Using Machine Learning Classification Algorithms.' Their study utilizes various machine learning methods, including Support Vector Machine (SVM), Random Forest (RF), Decision Tree (DT), Artificial Neural Network (ANN), and K-Nearest Neighbours (KNN) were employed for diabetes prediction. The research focuses on the early prediction and diagnosis of diabetes, a common global health concern. Notably, the random forest classifier demonstrated high accuracy in diabetes prediction, outperforming other machine learning approaches.

Alghamdi [11] sheds light on the critical role of data analysis and predictive techniques in addressing the complexities of diabetes, emphasizing the importance of early detection and effective management of this chronic disease. The study highlights the potential of machine learning algorithms, particularly the XGBoost classifier, for accurate diabetes prediction. XGBoost's gradient-boosting framework, tailored for large and intricate datasets, significantly contributes to enhancing diabetes prediction with an impressive accuracy rate of 89%. While XGBoost proves highly effective, the choice of the best algorithm may depend on specific data characteristics and research objectives. Moreover, the paper underscores that data analysis and predictive techniques extend beyond diabetes prediction, offering insights into risk factors, disease progression, treatment effectiveness, and underlying disease mechanisms. These methods hold promise for revolutionizing the timely identification and effective control of diabetes, a rapidly growing health concern.

The systematic review by Fregoso-Aparicio et al. [12] delves into the challenges surrounding the prediction of type 2 diabetes using machine learning and deep learning techniques. The study analysed 90 research works, employing PRISMA and Keele-Durham methodologies, and provides key insights. Heterogeneity among previous studies makes technique selection a challenge, while the lack of feature transparency reduces model interpretability. Among the 18 model types compared, tree-based algorithms outperformed deep neural networks. The incorporation of data balancing and feature selection techniques enhanced model efficiency, and the use of tidy datasets nearly yielded perfect model outcomes. This comprehensive review, featured in Diabetology & Metabolic Syndrome, offers valuable guidance for improved diabetes prediction.

[13] Study emphasizes the significance of data mining techniques (DMTs) in early disease prediction, focusing on Diabetes Disease (DD), a leading global cause of death. Using the Pima Indian Diabetes Data Set, the research employs SVM, Decision Stump, Naïve Net, and a Proposed Ensemble method (PEM) to predict symptoms. The PEM demonstrates high accuracy of 90.36%, highlighting its potential in diabetes prediction.

Evwiekpaefe et al. [14] delves into the pressing issue of Diabetes Mellitus (DM), a chronic disease with severe global implications. Using selected hospitals in Kaduna, the study creates a predictive model for diabetes, employing supervised learning algorithms like Decision Tree, K-Nearest Neighbor, and Artificial Neural Networks. The results indicate that Artificial Neural Networks achieved the highest accuracy at 97.40%, followed by the Decision Tree algorithm at 96.10%, and the K-Nearest Neighbor algorithm at 88.31%. This work highlights the potential of machine learning in addressing the diabetes epidemic.

3 Methodology

The methodology delineates the procedures integral to the diabetes prediction project through the utilization of logistic regression. It starts with importing necessary libraries data preparation, includes the core logistic regression training, and ends with model evaluation and accuracy assessment on both training and testing datasets. Figure 1 of block diagram visualizes the workflow and data flow throughout the project.

In our study on Diabetes Prediction using Logistic Regression, a crucial aspect was data preprocessing and visualization to ensure the quality of our predictive model. Preprocessing involved data cleaning, which included addressing missing values and outliers, standardizing features, and encoding categorical variables. Additionally, we conducted feature selection to pinpoint the most pertinent predictors for assessing diabetes risk.

To gain insights into the dataset, data visualization played a pivotal role. We employed various techniques such as histograms, box plots, and correlation matrices to visualize the distribution of key features and understand their relationships. These visualizations allowed us to identify patterns and potential correlations between variables, which informed the feature selection process. We also created interactive visualizations using Gradio, providing an intuitive platform for users to explore the data and model predictions.

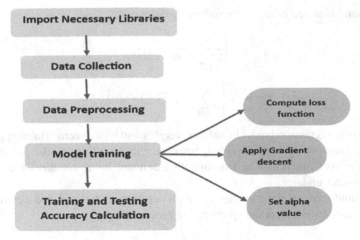

Fig. 1. Block diagram

By combining effective data preprocessing and visualization techniques, we enhanced the interpretability and accuracy of our Diabetes Prediction model, making it a valuable system for early diabetes risk assessment.

The methodology employed in this study aims to predict diabetes using logistic regression, a popular classification algorithm. The dataset is initially devided into training and testing subsets using a 70–30 train-test split ratio. This approach ensures both data independence and an efficient evaluation framework. The data is subsequently converted into NumPy arrays for efficient processing.

The essence of logistic regression lies in the sigmoid function, defined as:

$$Z = w.x + b \tag{1}$$

$$Sig(z) = \frac{1}{1 + e^{\wedge(z)}} \tag{2}$$

$$f_{\vec{w},b}(\vec{x}) = g(\vec{w} \cdot \vec{x} + b) = \frac{1}{1 + e^{-(\vec{w} \cdot \vec{x} + b)}} \tag{3}$$

where z denotes the linear combination of weights, features and the bias term, and sig(z) outputs the probability of the positive class.

To gauge the model's performance, we employ the logistic loss function, which measures the error between predicted values and actual labels.

$$J(\vec{w}, b) = -\frac{1}{m} \sum_{i=1} \left[y^{(i)} \log \left(f_{\vec{w},b}(\vec{x}^{(i)}) \right) + (1 - y^{(i)}) \log \left(1 - f_{\vec{w},b}(\vec{x}^{(i)}) \right) \right] \tag{4}$$

Here, y denotes the actual label, $f(x)$ is the predicted value, and the cost function measures dissimilarity among the predicted values by our algorithm and actual values.

The gradient descent function is defined, which iteratively updates the model parameters (weights and bias) using the computed gradients. This is the core of logistic regression training. The cost is also computed and stored at each iteration. We update values of weights and bias according to the following derivatives.

$$w_j = w_j - \alpha \left[\frac{1}{m} \sum_{i=1}^{m} (f_{\vec{\mathbf{w}},b}(\vec{\mathbf{x}}^{(i)}) - y^{(i)}) x_j^{(i)} \right] \tag{5}$$

$$b = b - \alpha \left[\frac{1}{m} \sum_{i=1}^{m} (f_{\vec{\mathbf{w}},b}(\vec{\mathbf{x}}^{(i)}) - y^{(i)}) \right] \tag{6}$$

The logistic regression model initializes weights and bias to zero. Training proceeds through the gradient descent algorithm, iteratively updating the model parameters to minimize the cost function. The learning rate, denoted by α, regulates the step size during parameter updates.

Throughout training, the cost values at each iteration are tracked, and a convergence plot is generated to visualize the optimization process (Fig. 2).

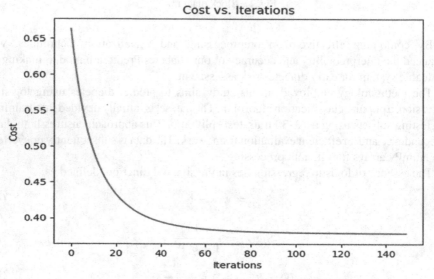

Fig. 2. Cost vs Iterations plot

For making predictions, the sigmoid output is computed using the trained model, and a binary classification is performed using a threshold, typically set at 0.5. Model accuracy is assessed by calculating the ratio of correctly predicted instances to the total original instances in the training and testing datasets, providing insights into the performance and generalization capability of the model.

3.1 Dataset Description

The required dataset was obtained from https://www.kaggle.com/datasets/iammustafatz/diabetes-prediction-dataset.

Diabetes dataset contains 1,00,000 cases. The dataset for diabetes prediction comprises medical and demographic data from patients, including their diabetes status (positive or negative). The data includes features such as age, hypertension, gender, heart disease, body mass index (BMI), HbA1c level (Hemoglobin level), smoking history, and blood glucose level. For training model we are using only 4 features- HbA1c level, blood glucose level, BMI and age.

We adopted a split of 70% for training and 30% for testing, effectively dividing the dataset of 100,000 records into a training subset of 70,000 and a testing subset of 30,000. This separation facilitated the development and comprehensive assessment of our machine learning model. Furthermore, Fig. 3 presents a pie chart illustrating the distribution of positive and normal cases within the dataset.

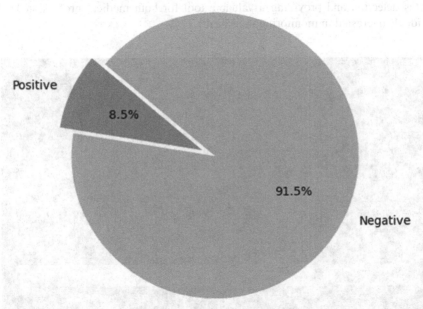

Fig. 3. Distribution of positive and negative cases

3.2 Technology Used

1. Python
2. Keras
3. Tensorflow
4. Jupyter Notebook
5. Matplotlib.

The project leverages several key technologies to accomplish its objectives. Python, a versatile and widely-used programming language, serves as the primary coding platform, providing a robust foundation for various tasks. To develop and fine-tune neural networks, the project utilizes tensorFlow, an open-source machine learning framework, alongside Keras, a high-level neural networks API. For interactive development, we employ Jupyter Notebook, providing a seamless environment for code execution and documentation. Visualization of data and model performance is facilitated by Matplotlib, a popular plotting library. These technology components collectively empower the project to effectively predict and analyze diabetes.

4 Results

We present the outcome of our Diabetes Prediction model using Logistic Regression integrated with a Gradio interactive web application. Gradio's user-friendly interface enhances accessibility, allowing users to input data effortlessly and obtain real-time predictions. This user-centric approach aligns with the objective of streamlining early diabetes detection and providing a valuable tool for both medical professionals and individuals interested in monitoring their health.

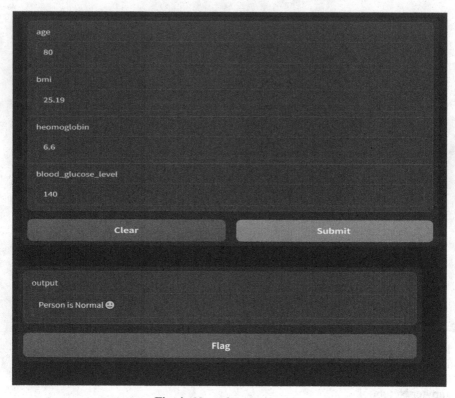

Fig. 4. Normal person's output

In our diabetes prediction model, we assessed the health status of individuals based on a set of crucial health parameters: Body Mass Index (BMI), age, hemoglobin level, and blood glucose level. These parameters were collected simultaneously, and the model provided a meaningful output that greatly aids in medical diagnosis. For the specific data point with features (age = 80 years, BMI = 25.19 kg/m^2, hemoglobin = 6.6 g/dL, and blood glucose level = 140 mg/dL) as shown in Fig. 4, our model conclusively determined the individual's health status as "Normal."

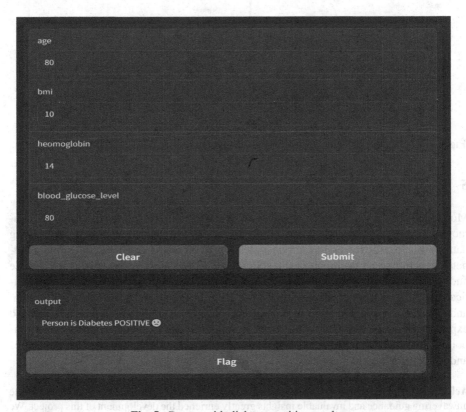

Fig. 5. Person with diabetes positive result

In Fig. 5 for the specific set of data with features (age = 80 years, BMI = 10 kg/m^2, hemoglobin = 14 g/dL, and blood glucose level = 80 mg/dL), our model unequivocally determined the individual's health status as "Diabetes Positive" (Fig. 6).

We conducted an insightful analysis of age distributions among individuals with and without diabetes. The visualizations depict these distributions vividly, revealing distinctive patterns. In the left figure, "Distribution of Age for People with Diabetes," the blue bars shows the age distribution for individuals with diabetes. On the right, the "Distribution of Age for People without Diabetes" displays the corresponding distribution for those without diabetes. These visualizations provide a clear understanding of the age demographics in both groups.

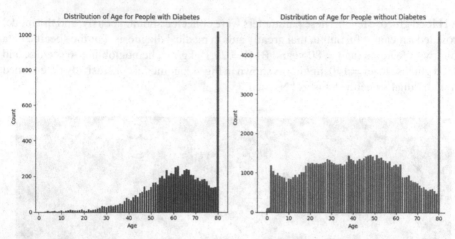

Fig. 6. Subplots depicting the distribution of age among individuals with and without diabetes.

5 Conclusion

Model achieved a remarkable validation accuracy of 91.51%, along with training accuracy of 91.49% highlighting its capability to understand and generalize from the training data. While logistic regression demonstrated its effectiveness as a predictive tool for diabetes, it is important to emphasize that clinical decisions must involve medical expertise and diagnostic tests. Research highlights the significance of proactive diabetes identification using machine learning techniques, setting the stage for ongoing exploration and the real-world application of healthcare solutions. With continued refinement and expanded datasets, our methodology can significantly aid healthcare practitioners inpipointing individuals who are at risk, potentially reducing the global burden of diabetes and improving patient outcomes.

Acknowledgement. Our heartfelt appreciation goes out to Prof. Zarinabegam Mundargi, whose unwavering guidance and invaluable insights greatly enriched the development of this project. We also extend our gratitude to Vishwakarma Institute of Technology for their generous provision of essential resources, enabling us to bring this project to life.

References

1. Cho, N., et al.: IDF diabetes atlas: global estimates of diabetes prevalence for 2017 and projections for 2045. Diabetes Res. Clin. Pr. **138**, 271–281 (2018)
2. Sanz, J.A., Galar, M., Jurio, A., Brugos, A., Pagola, M., Bustince, H.: Medical diagnosis of cardiovascular diseases using an interval-valued fuzzy rule-based classification system. Appl. Soft Comput. **20**, 103–111 (2014)
3. Varma, K.V., Rao, A.A., Lakshmi, T.S.M., Rao, P.N.: A computational intelligence approach for a better diagnosis of diabetic patients. Comput. Electr. Eng. **40**, 1758–1765 (2014)
4. Kandhasamy, J.P., Balamurali, S.: Performance analysis of classifier models to predict diabetes mellitus. Procedia Comput. Sci. **47**, 45–51 (2015). Appl. Sci. 2019, 9, 4604 16 of 18

5. Larabi-Marie-Sainte, S., Aburahmah, L., Almohaini, R., Saba, T.: Current techniques for diabetes prediction: review and case study. Appl. Sci. **9**, 4604 (2019). https://doi.org/10.3390/app9214604
6. Joshi, T., Pramila, M., Chawan, P.: Diabetes Prediction Using Machine Learning Techniques, pp. 2248–9622 (2018). https://doi.org/10.9790/9622-0801020913
7. Rani, K.M.: Diabetes prediction using machine learning. Int. J. Sci. Res. Comput. Sci. Eng. Inf. Technol. 294–305 (2020). https://doi.org/10.32628/CSEIT206463
8. Woldemichael, F.G., Menaria, S.: Prediction of diabetes using data mining techniques. In: 2018 2nd International Conference on Trends in Electronics and Informatics (ICOEI), Tirunelveli, India, pp. 414–418 (2018). https://doi.org/10.1109/ICOEI.2018.8553959
9. Bhat, S.S., Selvam, V., Ansari, G.A., Ansari, M.D., Rahman, M.H.: Prevalence and early prediction of diabetes using machine learning in North Kashmir: a case study of district Bandipora. Comput. Intell. Neurosci. **2022**, Article ID 2789760, 12 (2022). https://doi.org/10.1155/2022/2789760
10. Nahzat, S., Yağanoğlu, M.: Diabetes prediction using machine learning classification algorithms. Eur. J. Sci. Technol. **24**, 53–59 (2021)
11. Alghamdi, T.: Prediction of diabetes complications using computational intelligence techniques. Appl. Sci. **13**, 3030 (2023). https://doi.org/10.3390/app13053030
12. Fregoso-Aparicio, L., Noguez, J., Montesinos, L., García-García, J.: Machine learning and deep learning predictive models for type 2 diabetes: a systematic review. Diabetol. Metab. Syndr. **13**, 148 (2021). https://doi.org/10.1186/s13098-021-00767-9
13. Alehegn, M., Joshi, R., Mulay, P.: Analysis and prediction of diabetes mellitus using machine learning algorithm. Int. J. Pure Appl. Math. **118**, 871–878 (2018)
14. Evwiekpaefe, A.E., Abdulkadir, N.: A predictive model for diabetes using machine learning techniques (A Case Study of Some Selected Hospitals in Kaduna Metropolis) (2021)

Performance Analysis of Low Voltage Ride Through Techniques of DFIG Connected to Grid Using Soft Computing Techniques

Manohar Gangikunta[1](✉) [iD], Sonnati Venkateshwarlu[1] [iD], and Askani Jaya Laxmi[2] [iD]

[1] CVR College of Engineering, Hyderabad, India
manohar.gangikunta@gmail.com
[2] JNTUH, Hyderabad, India

Abstract. On demand with green generation in sustainable development. Renewable sources assure fascinating parameters for reduced operating cost along with increased life span. Technological developments in the domains of wind generators and turbines made the investors opt for wind energy generation. Varied speed with IGs is an attractive option with initial price independent watt less. Due to the advantage of harvesting huge amounts, Power system Operators (PSOs) to incline towards DFIGs. LVRT watt-less power using RFO & ANN controller and Grey Wolf Optimization (GWO) controller.

Keywords: Artificial Neural Network (ANN) · Doubly Fed Induction Generator (DFIG) · Low Voltage Ride Through (LVRT) · Random Forest Algorithm (RFO) and Grey Wolf Optimization (GWO) Algorithm

1 Introduction

Diminishing reserves of fossil fuels along with enhanced greenhouse gases has found a steep growth in terms of sustainable Development. Renewable energies are eminent encouraging, and their percentage of share was 12% in 2022 in global energy scenario. Nearly 7.3% of total energy generated in the world came exclusively from wind energy. European Country Denmark is leading with 721TWh 2021. India produced nearly 70TWh of wind energy in 2022 and its installed capacity increased to 46GW in 2023. With wind targeting get mounted original capacity.

Wind turbines producing fixed speed were deployed in initial years in Wind Energy conversion Systems (WECS) and more often used in stand-alone power systems. Later fixed speed wind turbine generator sets are tied to the grid. Basic advantage of such systems is that emf produced by such systems synchronize with grid frequency. These systems offer additional advantages of simple aerodynamic control and they are cost effective. However, they suffer from major drawback of requirement of capacitor banks for reactive power. With IG rotor based to control rotor resistance was implemented by a Danish manufacturer. With the width of pulse in semiconductor and hence induction generator. Speed regulated by this method is limited around 10% which is a major limitation of this method. Wattless power requirement associated with high power loss has

© The Author(s), under exclusive license to Springer Nature Switzerland AG 2024
S. L. Gundebommu et al. (Eds.): REGS 2023, CCIS 2081, pp. 62–72, 2024.
https://doi.org/10.1007/978-3-031-58607-1_5

method later technology. The abovementioned limitations are overcome by adjustable speed wind turbines. These turbines can align in the direction of the wind to maximise the speed and hence can generate high amount of power. Frequency of the induced voltages vary since emf is proportional to turbine speed. To connect such systems with electric grid, it is essential to have frequency converter which is realised by power electronic converters connected between rotor and grid. In such systems, the turbine rotor absorbs and minimize the mechanical fluctuations and hence power qual- ity concerns can be minimised. Different types of wind generators deployed in variable speed WECS are [1],

(i) Squirrel cage induction generators are a type of electrical generator that operate on the principle of electromagnetic induction. They are commonly used in various applications, such as wind turbines and industrial machinery. The design of these generators includes a rotor with conductive bars arranged in a squirrel cage shape, which allows for efficient power generation.

(ii) Slip ring induction generators, also known as wound rotor induction generators, are another type of electrical generator that utilize electromagnetic induction. Unlike squirrel cage generators, slip ring generators have a rotor with windings connected to external resistors through slip rings. This design enables better control over the generator's performance and allows for adjustable speed and torque characteristics.

(iii) Permanent magnet induction generators are a modern type of electrical generator that utilize permanent magnets instead of electromagnets to generate electricity. These generators offer several advantages, including higher efficiency, compact size, and reduced maintenance requirements. They are commonly used in renewable energy systems, such as wind turbines, due to their ability to efficiently convert mechanical energy into electrical energy.

The rating of the power converters used in adjustable speed WECS using Squirrel cage induction generators (SCIGs) ought to be same rating as that of induction generator. This makes the initial investment to be high and is major burden for utilities. Stator of the SCIG is tied directly to the grid. WECS employing squirrel cage genera- tors offers the advantages of high reliability along with low operating and maintenance costs. Additional advantage of such system is that the power is transmitted to the grid at unity power factor.

Permanent Magnet Synchronous Generators (PMSG) when used in WECS afford the advantages of low power loss along with simple control system. Additionally, power factor can be improved with PMSG. PMSG wind turbines are not suitable for bulk power generation due to non-availability of rare field earth magnets [2]. Further there are chances. Temperature control is achieved by colling system enhances the overall system.

With wound rotor is preferred IGs in turbines. For Bulk power generation Doubly Fed Induction Generators are widely deployed in most of the wind power parks and they are gaining more popularity in offshore wind.

2 Doubly Fed Induction Generator

These generators are became almost unanimous choice in variable speed wind energy systems when bulk energy need to be generated. Stator of this generator is directly coupled electric whereas rotor is coupled automatically. Since rotor speed wind turbine continuously varies, frequency of the generated voltage also varies. To match the frequency of generated voltage with grid frequency, power converters are connected in the rotor circuit. These converters act as frequency converter [3]. Power electronic converters need process only slip power Since this leads to reduced converter rating and hence initial cost of the DFIG is very much reduced. From Fig. 1, IG network.

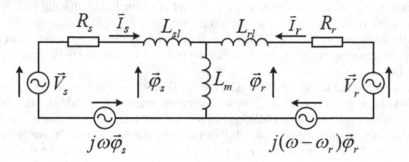

Fig. 1. IG schematic.

Rotating frame in synchronously is shown in Fig. 2 at ω [4].

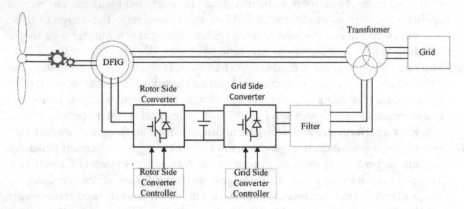

Fig. 2. IG Circuit diagram.

rotor and stator of Flux linkage vectors are given by

$$\vec{\Phi}_s = L_m \vec{I}_r + L_s \vec{I}_s \tag{1}$$

$$\vec{\Phi}_r = L_m \vec{I}_s + L_r \vec{I}_r \tag{2}$$

Where L_s is the magnetic inductance of stator $L_s = L_m + L_{sl}$ and
L_r is the magnetic inductance $L_r = L_m + L_{rl}$
L_{sl} and L_{rl} are leakages and L_m mutual

$$\vec{V}_s = R_s \vec{I}_s + \frac{d\vec{\Phi}_s}{dt} + j\omega \vec{\Phi}_s \qquad (3)$$

$$\vec{V}_r = R_s \vec{I}_r + \frac{d\vec{\Phi}_r}{dt} + j(\omega - \omega_r)\vec{\Phi}_r \qquad (4)$$

In rotating reference frame, stator voltage vector is described by the equation.

$$v_{sd} = R_s i_{sd} + \frac{d\Phi_{sd}}{dt} - \omega \Phi_{sq} \qquad (5)$$

$$v_{sq} = R_s i_{sq} + \frac{d\Phi_{sq}}{dt} - \omega \Phi_{sd} \qquad (6)$$

Where i_{sd} and i_{sq} are the components of stator current in d_s - q_s synchronous frame and v_{sd} and v_{sq} are stator d axis and stator q axis components.

Stator useful power and wattless power of a DFIG when implemented in field-oriented control are given by

$$P_s = \frac{3}{2}\frac{L_m}{L_s}\omega_e \, \Phi_{sd} i_{rq} \qquad (7)$$

$$Q_s = \frac{3}{2}\frac{L_m}{L_s}\omega_e \Phi_{sd}\left(\frac{\Phi_{sd}}{L_m} - i_{rq}\right) \qquad (8)$$

3 LV Ride Through

The stator gets short circuited and its magnitudes equal to zero, in terms of grid faults. In bulk generators, drop in the stationery part of IG is considerably low when compared to stator voltage and can be neglected.

$$\vec{V}_s^s = R_s \vec{I}_s^s + \frac{d(\vec{\psi}_s^s)}{dt} \qquad (9)$$

From above equation if stator drop is neglected, we can conclude that stator RMF is constant. Stator magnetic field is held stationery and rotor magnetic field is rotating at ω_r, difference in speed is $(1-s)\omega\_s$, which is around 30%. Generator with 100 MW, power 30 MW with dead short circuit the rotor power is raised to 70 MW. This increased rotor power in circuit. Circuit fails if once it saturates. In initial, Renewable Energy Conversion Systems used to disconnect from the grid as soon as short circuit is detected. WECS may still make the main supply worse [5]. Crowbar circuits were used to treat such type of problems The drawback of crowbar circuit is that when it activates, the generator draws reactive power from grid and SQIM. New grid codes, implemented by several countries require wind parks may be interfaced to main source, and inject which is not watt [6–7].

The capability of the machine to connect to main supply during supply disturbances or magnitude drops can be referred to as Ride through (LV). Figure 3 mentions the protection circuit, which serves from fault, the event with sinks. Consideration of case of grid errors, protection circuit initiated, rotor with voltage is zero with resistors. These resistors minimize the rotor current to safeguard the power converters. The major restraint of this method is that when protection circuit initiated, generator acts SQIM and draws bulk magnitudes of wattless power. According to existing case faults, wind parks need to impend to main source. This limitation of protection circuit initiated triggered to search for alternative methods to enhance ride through capability of generators.

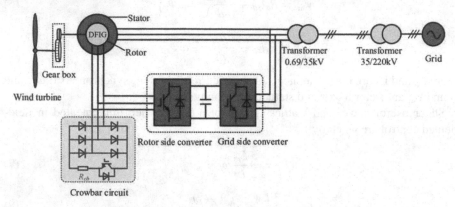

Fig. 3. DFIG protection circuit using a crowbar mechanism.

Different literatures presents various LVRT techniques, which are visually represented in Fig. 5 as a flow chart.

By connecting a chopper in DC link LVRT capability can be enhanced [8]. By regulating duty ratio of chopper circuit, magnitude is controlled to be same value, hence converters protected during grid faults. Protection by chopper is depicted in Fig. 4. Unnecessary wastage of power in rotor resistors along with requirement of high choppers make this method absolute in subsequent years.

Storage systems embedded in DC link of the rotor system has potency to regulate the DC link voltage and protect the converters during grid errors [9–10]. Requirement of high rating rotor side converter and complexity in circuit are major limitations of this method. ESS method to improve fault ride through shown in Fig. 6.

4   Results and Simulation

Low voltage ride through schemes can be categorized into three categories: extra identities combined with different controller techniques [11–15]. The use of additional auxiliaries' techniques leads to increased costs, area also into consideration. However, when are extra identities combined designs, it enhances the technical performance at

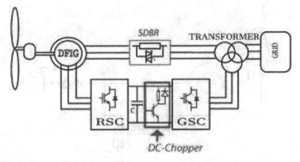

Fig. 4. A chopper circuit is employed to reduce disturbances

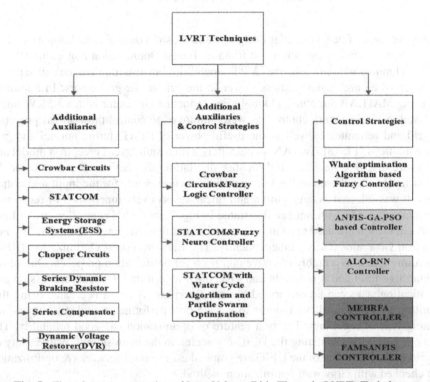

Fig. 5. Flowchart representation of Low Voltage Ride Through (LVRT) Techniques.

the expense of additional cost. Many researchers proposed the use of artificial intelligent controllers to achieve LVRT enhancement of DFIG without the need for auxiliaries. Another approach suggested by C. J. Mohan and N. Kumar [16] is the utilization of power [17] implemented and innovative SMC utilizing technique for Low voltage ride through enhancement of IG [18] adopted a Genetic algorithm based LQR controller for LVRT enrichment of DFIG. H Ahmadi et al [19] proposed the use of a Multi-objective Krill algorithm for LVRT enrichment during grid faults. These advancements have inspired

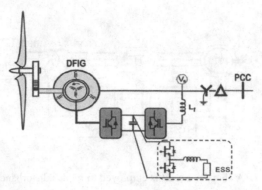

Fig. 6. Enhancement of LVRT of DFIG using ESS

the development of new controller designs based on soft computing techniques such as gray wolf optimization algorithm and Random Forest Optimization Algorithm (RFO). These techniques not only enhance LVRT capabilities but also improve the active power output of DFIG and inject wattless power to the grid under grid errors. In a simulation using MATLAB Simulink, a Doubly Fed Induction Generator with a 3.5KW rating was used. Neural network controllers were employed to control the wattless power of the grid and generator, as well as the useful power of DFIG, during normal and grid fault conditions [17, 18]. Two ANN controllers with a multilayer Perceptron model and a structure of 2–7-1 were selected for this simulation. The activation function for the hidden layers was chosen as the Log sigmoid function, while for the input and output layers, it was selected as tan sigmoid and linear, respectively functions. For regulating the necessary power with wattage illustrated In Fig. 7, two ANN controllers are utilized in this simulation are used to Manage the active and reactive power of the Doubly Fed Induction Generator (DFIG) independently. The Random Forest Optimization (RFO) algorithm represents a highly effective approach for optimization purposes. One of the popular algorithm and it is simple and diversified algorithm [20]. Both regression and classification tasks can be performed by this algorithm. A forest is created using this algorithm by using available models to get the best performance and it is supervised learning type of algorithm. The best feature of optimization has good reliability. The RFO employed for simulating the DFIG connected to the main supply. Wattless qty of the grid the useful power of the DFIG are graphed using neural networks & optimization then checked with Gray wolf optimization method.

Grey Wolf (GWO) optimization is a nature inspired, meta heuristic al- gorithm proposed by Mirzalili et all in 2014 [21, 22].It imitates the hunting behavior of stack of grey wolves. The same behavior is used in optimizing the DFIG network. GWO algorithm uses different types of recurrences, namely α(alpha),β(beta),delta(ω)and omega(ω).The best solution for a given problem is given by alpha wolves and they act as leaders of their group [23–26]. The second in hierarchy is beta wolves. Beta wolves obey the instructions from alpha wolves and instruct delta and omega wolves. The hierarchical behavior is shown in Fig. 8 (Figs. 9, 10 and 11 see Table 1).

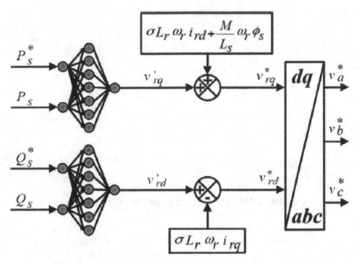

Fig. 7. DFIG with ANN to control different powers

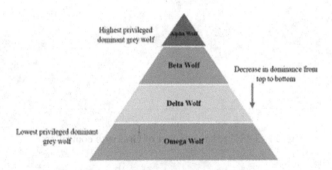

Fig. 8. Hierarchy of wolves in grey wolf optimization

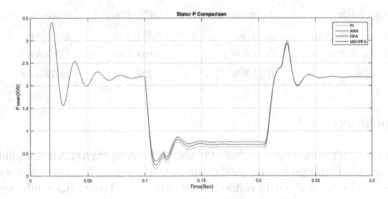

Fig. 9. Active power with control algorithms

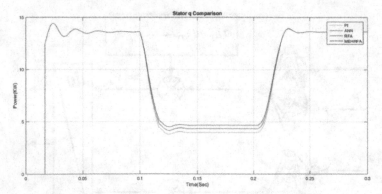

Fig. 10. DFIG Stator Kvar power with controllers

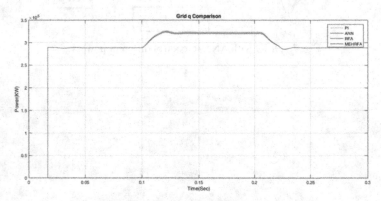

Fig. 11. DFIG reactive power of Grid with controllers

Table 1. Grid and DFIG comparison of power

Type of Controller	Necessary power	DFIG Power	Reactive Power
Proportional + Integral	0.7 KW	0.55KVAr	28KVAr
Artificial Neural Network	0.74 KW	0.65KVAr	27KVAr
RFO	0.77 KW	0.75KVAr	26.2KVAr

5   Conclusion

Reasons for LVRT solutions with dual IG explained clearly. MATLAB Simulink was interfacing DFIG with grid using, ANN control technique, RFO technique grey wolf optimization algorithm. From the results, observed that GWO controller exhibit superior performance compared to ANN and RFO controllers in terms of active and wattless powers during grid faults .

References

1. Harish, V.S.K.V., Sant, A.V.: Grid integration of wind energy conversion systems. In: Alternative Energy Resources: The Way to a Sustainable Modern Society, pp.45–66 (2021)
2. Tripathi, S.M., Tiwari, A.N., Singh, D.: Grid-integrated permanent magnet synchro- nous generator based wind energy conversion systems: a technology review. Renew. Sustain. Energy Rev. **51**, 1288–1305 (2015)
3. Karad, S., Thakur, R.: Recent trends of control strategies for doubly fed induction generator based wind turbine systems: a comparative review. Arch. Comput. Methods Eng. **28**, 15–29 (2021)
4. Hu, J.B., He, Y.K.: Dynamic modelling and robust current control of wind-turbine driven DFIG during external AC voltage dip. J. Zhejiang Univ.-Sci. A **7**(10), 1757–1764 (2006)
5. Ezzat, M., Benbouzid, M., Muyeen, S.M., Harnefors, L.: Low-voltage ride- through techniques for DFIG-based wind turbines: state-of-the-art review and future trends. In: IECON 2013–39th Annual Conference of the IEEE Industrial Electronics Society, pp. 7681–7686. IEEE (2013)
6. Singh, B., Singh, S.N.: Wind power interconnection into the power system: a review of grid code requirements. Electr. J. **22**(5), 54–63 (2009)
7. Swain, S., Ray, P.K.: Short circuit fault analysis in a grid connected DFIG based wind energy system with active crowbar protection circuit for ride through capability and power quality improvement. Int. J. Electr. Power Energy Syst. **84**, 64–75 (2017)
8. Wang, M., Xu, W., Hongjie, J., Yu, X.: A new control system to strengthen the LVRT capacity of DFIG based on both crowbar and DC chopper circuits. In: IEEE PES Innovative Smart Grid Technologies, pp. 1–6. IEEE (2012)
9. Jin, C., Wang, P.: Enhancement of low voltage ride-through capability for wind turbine driven DFIG with active crowbar and battery energy storage system. In: IEEE PES General Meeting (pp. 1–8). IEEE. Author, F.: Article title. J. **2**(5), 99–110 (2016)
10. Guo, W., Xiao, L., Dai, S.: Enhancing low-voltage ride-through capability and smoothing output power of DFIG with a superconducting fault-current limiter–magnetic energy storage system. IEEE Trans. Energy Convers. **27**(2), 277–295 (2012)
11. Manohar, G., Venkateshwarlu, S., JayaLaxmi, A.: An elite approach for enhancement of LVRT in doubly fed induction generator (DFIG)-based wind energy conversion system (WECS): a FAMSANFIS approach. Soft. Comput. **26**(21), 11315–11337 (2022)
12. Manohar, G., Venkateshwarlu, S., Laxmi, A.J.: Enhancement of low voltage ride through capability of doubly fed induction generator based wind turbine using fuzzy logic controller. In: E3S Web of Conferences, vol. 309, p. 01047. EDP Sciences (2021)
13. Rashid, G., Ali, M.H.: Nonlinear control-based modified BFCL for LVRT capacity enhancement of DFIG-based wind farm. IEEE Trans. Energy Convers. **32**(1), 284–295 (2016)
14. Obulesu, D.: Performance Improvement of Grid-Connected DFIG-Based Wind Turbine with a Fuzzy-Based LVRT Controller. Turk. J. Comput. Math. Educ. (TURCOMAT) **12**(6), 3599–3605 (2021)
15. Alsmadi, Y.M., et al.: Detailed investigation and performance improvement of the dynamic behavior of grid-connected DFIG-based wind turbines under LVRT conditions. IEEE Trans. Ind. Appl. **54**(5), 4795–4812 (2018)
16. Aguilar, M.E.B., Coury, D.V., Reginatto, R., Monaro, R.M.: Multi-objective PSO ap- plied to PI control of DFIG wind turbine under electrical fault conditions. Electr. Power Syst. Res. **180**, 106081 (2020)
17. Zheng, X., Chen, X.: Enhancement on transient stability of LVRT of DFIG based on neural network D-STATCOM and crowbar. In: 2017 11th IEEE International Conference on Anti-counterfeiting, Security, and Identification (ASID), pp. 64–68. IEEE (2017)

18. Puliyadi Kubendran, A.K., Ashok Kumar, L.: LVRT Capability improvement in a grid- connected DFIG wind turbine system using neural network-based dynamic voltage restorer. In: Kumar, L., Jayashree, L., Manimegalai, R. (eds.) Proceedings of International Conference on Artificial Intelligence, Smart Grid and Smart City Applications. AISGSC 2019 2019, pp. 11–19. Springer, Cham (2020). https://doi.org/10.1007/978-3-030-24051-6_2

19. https://www.javatpoint.com/machine-learning-random-forest-algorithm/lncs. Accessed 02 Sep 2023

20. Abdelateef Mostafa, M., El-Hay, E.A., ELkholy, M.M.: Recent Trends in wind energy conversion system with grid integration based on soft computing methods: comprehensive review, comparisons and insights. Arch. Comput. Methods Eng. **30**(3), 1439–1478 (2023)

21. Syed, S.N., Kalyani, S.T.: Performance improvement of doubly fed induction generator using grey wolf optimization. In: 2021 International Conference on Advances in Electrical, Computing, Communication and Sustainable Technologies (ICAECT), pp. 1–7. IEEE (2021)

22. Mirjalili, S., Mirjalili, S.M., Lewis, A.: Grey wolf optimizer. Adv. Eng. Softw. **69**, 46–61 (2014)

23. Swarupa,L., Lakshmi, S., Rubanenko, O., Danylchenko, D.: Modeling, simulation and simultaneous tuning employing genetic algorithm in power system with power system stabilizer with TCSC controller. In: 2022 IEEE 3rd KhPI Week on Advanced Technology (KhPIWeek), Kharkiv, Ukraine, pp. 1-5 (2022).https://doi.org/10.1109/KhPIWeek57572.2022.9916504

24. Lakshmi,G.S., Harivardhagni, S., Divya, G., Lavanya, V.: Solar and bio-mass based hybrid power system for rural areas. In: 2020 IEEE-HYDCON, Hyderabad, India, pp. 1-5 (2020).https://doi.org/10.1109/HYDCON48903.2020.9242786

25. Bhargavi, R.N., Swarupa, M.L., Rajitha, M.: Power transformer protection using ANN and wavelet transforms. In: 2021 7th International Conference on Advanced Computing and Communication Systems (ICACCS), Coimbatore, India, pp. 161–166 (2021). https://doi.org/10.1109/ICACCS51430.2021.9441828

26. Lokeshwar Reddy, C., Sree Lakshmi, G.: Design of cascaded multilevel inverter-based STATCOM for reactive power control with different novel PWM algorithms. In: Sengodan, T., Murugappan, M., Misra, S. (eds.) Advances in Electrical and Computer Technologies. ICAECT 2021. Lecture Notes in Electrical Engineering, vol. 881. Springer, Singapore (2021) https://doi.org/10.1007/978-981-19-1111-8_80

Modelling and Methods of Green Computing

Passive Islanding Detection and Load Shedding Techniques in Micro Grids: A Brief Review

Sareddy Venkata Rami Reddy[1]($\boxtimes$) ⃝iD, T. R. Premila[1] ⃝iD, and Ch. Rami Reddy[2] ⃝iD

[1] Electrical and Electronics Engineering, Vels Institute of Science, Technology and Advanced Studies, Chennai 600117, India
svrami@gmail.com, premila.se@velsuniv.ac.in
[2] Electrical and Electronics Engineering, Joginpally B R Engineering College, Hyderabad 500075, India

Abstract. The passive islanding detection methodologies for the integrated DG approach are the primary subject of this study. Combined with fossil fuel sources and utilized in a cumulative manner, renewable energy generation and the grid are helping to satisfy the growing load demand. Unintentional islanding is the primary challenge of renewable arrangement coordination with grid connectivity. In the event of energy system islanding, the DG will detach from the main grid and begin supplying electricity to locally linked loads. According to many interconnections of DG systems, islanding will be recognized within 2 s using the techniques now in use for isolating the DG hardware from the grid. 'Remote controlling procedures' refer to islanding techniques that operate from the utility side, whereas 'local controlling procedures' are those that are implemented on the DG side. The methods for detecting passive islanding and the characteristics by which they may be evaluated are the topic of this study. The study provides a thorough discussion of the pros and cons, energy concerns, Zone of Non-Detection, islanding detection of balanced state, and merits of many different passive islanding detection approaches. For researchers developing advanced islanding detection systems, this paper's thorough explanation of Passive islanding strategies is a valuable resource.

Keywords: Passive islanding detection · Load shedding · Micro Grids · Distributed Generation (DG) · Local islanding · non-detection zone

1 Introduction

The global power grid is transitioning towards a renewable energy system based on DG to fulfil the growing demand for electricity. Renewable energy has several advantages, including a lower environmental impact and lower generating costs, which has led to a shift in customer preferences. Energy is created on the consumer side in the renewable DG based power system, and users may both receive power from and contribute power to the grid [2]. Small-scale power production describes this setup which uses renewable sources like the wind, the sun, and the tides to create a negligible quantity of electricity. Unintentional islanding is the primary challenge of renewable energy systems [3]. Islanding was shown in Fig. 1. The smart grid's circuit breakers, transformer,

© The Author(s), under exclusive license to Springer Nature Switzerland AG 2024

S. L. Gundebommu et al. (Eds.): REGS 2023, CCIS 2081, pp. 75–88, 2024.
https://doi.org/10.1007/978-3-031-58607-1_6

and required metres were all linked by this technology. We may disconnect the circuit breaker from the smart grid whenever we need to protect the local load and our renewable energy producing system. Figure 2 illustrates the differences between distant and close island detecting techniques. Furthermore, local islanding detection methods may be further categorised into passive, active, and hybrid modes [9]. Constraints on sympathetic islanding may be roughly divided into two categories: NDZ and electrical distributions interconnection characteristics. The Non-Detective Zone [10, 11] is the region where no section islanding technique could locate any islands. The preponderance of dispersed generating issues is a direct result of the connectivity quality of the electrical distribution network. Protecting the integrity of this power system requires optimisation algorithms, energy management methods, and monitoring of islanding grids. The benefits and drawbacks of the various islanding detection methods were outlined in detail in the study. More study is needed to enhance the various technologies of future islanding detecting techniques.

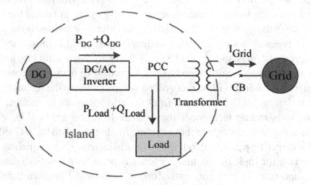

Fig. 1. Islanding in the power system

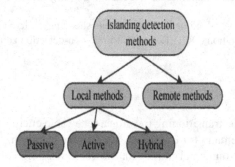

Fig. 2. Types of technics in islanding detection

2 Passive Islanding Detection (PID) Techniques

Measuring potential, current flow, active power, frequency, quality Power, phase angle, total harmonic distortion and harmonics at shared connections is ideal for this method of island detection. While small shifts in passive parameters are favored while the DG is operating in grid-connected mode, large shifts in these variables are employed to detect islanding [12]. In Fig. 3 we see the steps involved in the identification of PID systems. Switching disturbances, short circuit failures, and so on all have their respective islanding and non-islanding activities correctly handled by the threshold parameters. This method is the most accurate and straightforward to implement, but its main drawback is its enormous NDZ [13, 14].

Table 1. Different quality range of Island detection parameters [4–8]

Standard Factor	Quality (ms)	Time Detection	Frequency (Hz) Range	Range of Voltage
IEEE 1547	1	T<2000	$59.3 \leq f \leq 60.5$	$88\% \leq V \leq 110\%$
IEC 62116	1	T<200	$f0-1.5 \leq f \leq f0+1.5$	$85\% \leq V \leq 115\%$
Korean Standard	1	T<500	$59.3 \leq f \leq 60.5$	$88\% \leq V \leq 110\%$
UL 1741	≤ 1.8	T<2000	setting value	setting value
VDE0				
126-1-1	2	T<200	$47.5 \leq f \leq 50.2$	$80\% \leq V \leq 115\%$
IEEE 929-2000	2.5	T<2000	$59.3 \leq f \leq 60.5$	$88\% \leq V \leq 110\%$
AS47773 - 2005	1	T<2000	setting value	setting value

2.1 Over/Under Frequency and Voltage Method

The OUV and OUF are the first major methods used. By examining the actual threshold parameter of a voltage and frequency at a common coupling point where voltage/frequency within constrained parameters is evaluated from Eqs. (1), (2) described below, this OUV or OUF technique commonly used to determine whether or not islanding is necessary.

$$\left(\frac{V}{V_{max}}\right)^2 - 1 \leq \frac{\Delta P}{P_{DG}} \leq \left(\frac{V}{V_{min}}\right)^2 - 1 \tag{1}$$

$$Q_f\left(1-\left(\frac{f}{f_{min}}\right)^2\right) \leq \frac{\Delta Q}{P_{DG}} \leq Q_f\left(1-\left(\frac{f}{f_{max}}\right)^2\right) \tag{2}$$

Whereas V_{Max}: maximum voltage

V_{Min}: minimum voltage

f_{Max}: maximum frequency at the threshold

f_{Min}: minimum frequency at the threshold

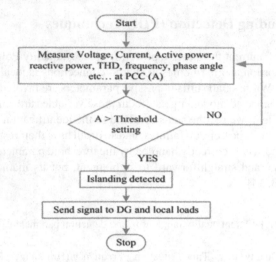

Fig. 3. PID techniques flow chart

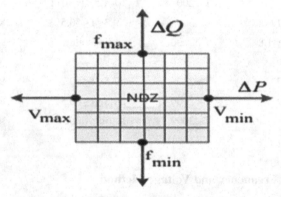

Fig. 4. PID technique in NDZ of OUV/OUF of NDZ

Table-1 shows that the minimum and maximum frequencies, V_{min} and f_{min}, are 88% and 59.5 Hz, respectively [15–17]. When the values of V and f exceed or fall below predetermined thresholds, islanding is detected. Figure 4 depicts the non-detection zone (NDZ) for this approach, which is the failure in islanding detection within the zone [49–51].

2.2 Rate of Change of Frequency

By calculating the value frequency at the point of common coupling, this article describes the Rate of change of frequency (ROCOF) method. The frequency of a system can be determined by using a PLL. In the event of a grid failure, the system's load will shift, causing frequency variations at the output, a phenomenon known as ROCOF [18, 22, 47, 48]. The frequency interval [7, 19] that corresponds to the ROCOF Eq. (3) is listed below.

$$\frac{df}{dt}(k) = \frac{f(t_k) - f(t_k - \Delta t)}{\Delta t} \qquad (3)$$

where the frequency f (t_k) at a time of kth example, f $(t_k - \Delta t)$ is the value of frequency Δt behind the Kth section time i.e., t_k-Δt. This method fails to detect level islanding and has a small NDZ compared to the passive OUV and OUF methods previously used (Refs. [4, 5]). Unravelling islanding and switching actions in these techniques relies heavily on the landscape of threshold parameters [8].

2.3 Rate of Change of Active and Reactive Power

Its primary function is to identify islanding. When compared to the baseline condition of being linked to the grid under normal operating conditions, the electrical metrics indicating ROCOP changes are greater during an islanding situation [20]. For the sake of safety, it is important to think about the right threshold while isolating switching processes in islanding procedures [21]. The method is best suited for the problem of steady islanding in which imbalanced loads have not been detected [40–46].

2.4 Voltage Unbalance (vu)

At the Point of Common Coupling, the VU of the three-phase voltages is determined via the passive islanding method. Sometimes VU experiences slight variations in load when power outages occur on the grid. For island detection, the Negative Sequence Voltage (NSV) to Normal Sequence Voltage (NSV) ratio is chosen as the VU Factor [22, 23]. In Expression (4), the VUF is defined as

$$VU_t = \frac{NSV}{PSV} \times 100 \qquad (4)$$

where VUt is the VU at the instant of time 't'. This expression (5) will represent, the change of VU from steady value of voltage at the time of load change and islanding.

$$VU_t = \frac{VU_s - VU_t}{VU_s} \qquad (5)$$

where VUs is the voltage imbalance of a predefined rate and VUt is the voltage imbalance at t^{th} point of time.

2.5 Phase Jump Detection

The islanding situation will be found by the phase angle detection method if the difference in voltage and current phase between the inverter outputs is greater than the basic threshold value [25]. A tweaked version of the PLL system is used to measure the phase angle difference. The swings of inverter voltage in the path and current at the moment islanding occurs, as can be seen in Fig. 5, are what causes the problem. The F fault happens because there is a new track of voltage being output by the inverter. Suppose the error deviation is high in comparison to the threshold parameter; in this case, islanding has been established, and the inverter has been forced to trip. A methodology's benefits include improving power quality and not affecting transient behaviour [24].

2.6 Voltage Unbalance and THD-Based Islanding Detection

It requires an islanding period when VU and current THD variances exceed a pre-set threshold. There are minor harmonics present in the grid-linked structure, amounting to less than 5% [25, 50]. The inverter will be linked to the load in the event that the scheme is islanded. The harmonic component of voltage and current increases as non-linear loads switch and load impedance varies. Using the VU approach [23] grouping in conjunction with the THD method is what is done in order to get rid of the false tripping that is intended for islanding operations. [26–28], it is together ROCORP and THD that are having more than pre-set criteria. The most useful method for detecting islanding is to compare the rate of change of reactive power, or ROCORP, with the total harmonic distortion, or THD.

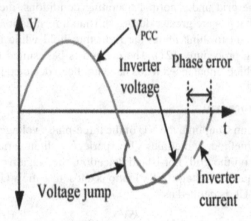

Fig. 5. Operation of PJD

2.7 Rate of Change of Frequency Over Active Power

The ROCOF over active power (df/dp) approach is greatly aided by this technique, making it ideal for identifying islanding. If you compare this approach to the ROCOF relay, you'll see that its NDZ is much smaller [29–39]. Additionally, it may identify low-to medium-severity power islanding incidents. Since ROCOF and ROCOAP settings are needed, it is challenging to fix the pre-set parameters to this technique.

3 Signal Processing Methods

3.1 S-Transform (ST)

The ST is defined as growth in the WT that incorporates phasor data. When operating at flexible and extensible narrow Gaussian windows, the S-Transform is employed for islanding identification and synthesis of power-related metrics. In the presence of noise and multi-resolution approaches, ST assisted in obtaining accurate results by analyzing each phase frequency separately. NSV processing based on the ST is described, if the spectral energy associated with a S outline is located and estimated after the current signal has been received. Formulating the energy terms and the primary change in the contour allows for the detection of islanding. Dimming with the ST technique helps lessen power quality problems during transients.

3.2 Wavelet Transform (WT)

It's the best tool for enhancing the quality of your voice or wave. The wavelet transform, using PCC voltage signals as input, passively finds the islanding problem. Using a Q-factor of 2.5 for parallel Resistance-Inductive-Capacitive loads, this method will yield the greatest results in inverter-reliant distribution generating systems. If the load's Q-factor is less than 2.5, power quality issues are unaffected. When a power signal arrives at a node via ROCOP, the wavelet packet transform can determine the islanding condition closest to the switching actions by analyzing the rate of change at the node. The spectral fluctuation in the high components at the voltage variation at PCC is utilized to detect islanding when the NDZ is close to zero and the cycle period is less than 2.5 cycles in a few specific cases.

3.3 Fuzzy and S- Transform

This method is best suited for islanding identification since it uses two distinct methods: the quick S-transform and the fuzzy rule method. Consequently, the outputs of the goal distributed generation's NSC and NSV are sent into the S-transform. Therefore, fuzzy rule dependently separates the islanding from the non-islanding behaviors. This has been proven for a wide variety of DG systems, and the islanding detection was achieved in a single cycle [7]. This method clearly distinguishes between switching and islanding behaviors, even in the face of significant disruptions.

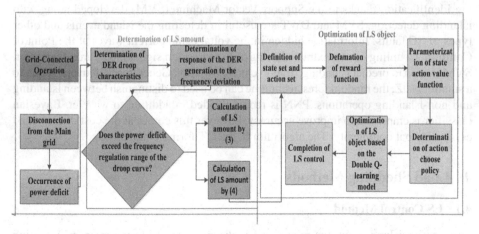

Fig. 6. LS control scheme.

3.4   Discrete Wavelet Transform (DWT) and S Transform

To accomplish islanding detection utilizing the S-transform and DWT techniques, it is necessary to first rip out NSV from DG sources. NSV then detects the islanding in numerous cases using S-transforms and wavelets. The nature of a distributed generating system can be determined in addition to the sort of islanding behavior being detected. After processing a voltage signal on PCC, power quality issues and islanding procedures are separated out into their own distinct groups. Current signals at PCC at 60 Hz with almost balanced islanding in the 1/3rd cycle (5.5 ms) are used to isolate individuals employing DWT islanding.

3.5   Fast Gauss Newton Algorithm (FGNA)

By extracting the characteristics of FGNW, a novel component is modelled that can tell the difference between islanding and non-islanding behaviors. The latest FGNW algorithm for evaluating a recursive also in a decoupled way, as whole current & voltage signal parameters precisely for accurate power systems despite significant noise is derived by minimizing the weighted error cost function and estimating the resulting hessian matrix by removing the off-diagonal parameters. With this strategy, islanding can be identified after just one cycle.

3.6   Artificial Intelligent Techniques (AIT)

Artificially intelligent methods such as PNNs (Probabilistic Neural Networks), SVMs (Support Vector Machines), ANNs (Artificial Neural Networks), and DTs (Decision Trees) are used to identify islanding once features have been extracted via signal refining procedures. Although these methods make calculations more complicated, the tool's advantage is that it can reduce the NDZ to zero. Voltage and current characteristics, as opposed to active, reactive powers and frequency, are now used exclusively for islanding identification thanks to the Support Vector Machine (SVM) developed along with islanding detection. SVM and DWT are useful in detecting the island events and other types of SC faults. Information hidden in the voltage and current waves at the Point of Common Coupling can be extracted using auto regression signal modelling; thereafter, SVM is used to predict islanding. To prevent Distribution Generator fault tripping due to a smaller NDZ, the random forest technique can be used to distinguish between islanding and non-islanding operations. PNN is recommended in addition to wavelet. Bayesian classifier is employed to improve accuracy, however this comes at the cost of increased computational complexity. The algorithm in the DT form is implemented.

4   Load Shedding Methods

4.1   LS Control Method

In a passive island setup, the micro grid is disconnected from the main grid, cutting off access to the main grid and the external power source. The frequency fluctuations in micro grids were a result of the Distributed Energy Resources' (DER) inability to meet load requirements. Improvements in DER generation capacity has resulted in a less active power imbalance. Two issues are presented in Fig. 6 that must be addressed by the LS control.

A micro grid when a bidirectional islanding event occurs in a micro grid, the system reacts quickly. Many problems will emerge after the system is set up with an LS control strategy. The LS control mechanism lessens the likelihood of such systemic occurrences.

During the LS event, we will be taught how to minimise loads and keep key loads running without interruption.

The major proposed LS control scheme steps are as follows:

✓ The mains will be disconnected from the micro grid automatically if a failure occurs in the mains.
✓ The correlation between active and frequency variation in a micro grid establishes LS parameters.
✓ Double Q-learning will be generated based on the load priority. The primary cargo is shipment.

4.2   LS Control Method Depends on Double-Q Study

The Double Q-learning method based on the LS scheme will be used to address the challenges of micro grid is landing and the power system shortfall. This article covered the full LS scheme, from its section on load removal through its discussion of critical load and critical load handling to its discussion of frequency stability (Fig. 7).

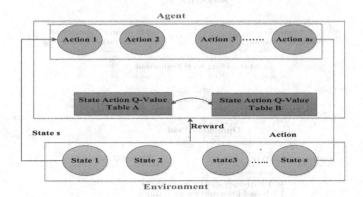

Fig.7. Block diagram of a Double-Q learning

4.3   DLC Approach

Various steps involved in the DLC approach are as follows:

✓ If the shortage and active deficiency happened as a result of MG islanding, then it is imperative to regularly monitor the frequency.

✓ An LSL is defined and calculated locally based on data collected by agents and on the needs of customers who have planned loads and capacities on the system.
✓ We shall optimise the distribution used and practically attained, with DCLS reaching consensus among LSLs based on the sub gradient algorithm of MAS.
✓ The DCLS will only be used to update local data, and the properties of a FID response changed into GID into DLF eccentricity will be made aware.
✓ At the point that LSLs determine that a system's frequency has returned to its rated value, the DCLS is activated, and MG islanding resolves to operate as a new state.

4.4 Different SCADA Methods

The implemented islanding on the WAMS contains the subsequent categorizations as depicted in Fig. 9.

Locate the hierarchical clustering procedure that takes into account the various islanding scenarios in the system.

By tallying all the instances of island-hopping, founded on best-case scenario selection and supplementary objective functions (Fig. 8).

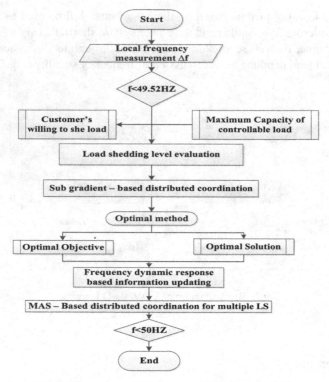

Fig. 8. Flow chart of DLC approach for load shedding

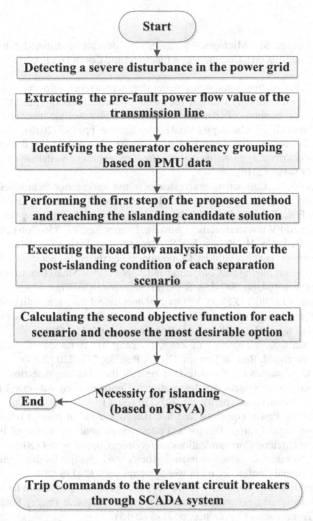

Fig. 9. The flowchart of SCADA system for load shedding

5 Conclusion

New passive islanding and load-shedding techniques are briefly reviewed in this article. The investigation revealed that customer islanding poses a threat to their safety and their ability to maintain connections. This necessitates a detection time of shorter than 2 s, after which DG must be disabled. Low power mismatch scenarios between the DG and the islanded load are not being detected by the passive approaches. After islanding, stabilizing the DG and associated load in the islanded area is best accomplished through the use of load-shedding mechanisms.

References

1. Zhang, S., Huang, M.: Microgrid: a strategy to develop distributed renewable energy resources. In: International Conference on Electrical and Control Engineering, pp. 3520–3523 (2011)
2. Elangovan, S.: Recent trends in sustainable development of renewable energy. In: International Conference on Advances in Electrical Technology for Green Energy, pp. 148–150 (2017)
3. Tran, Q.-T.: New methods of islanding detection for photovoltaic inverters. In: IEEE PES Innovative Smart Grid Technologies Conference Europe, pp. 1–5 (2016)
4. Khamis, A., Shareef, H., Bizkevelci, E., Khatib, T.: A review of islanding detection techniques for renewable distributed generation systems. In: Renewable and Sustainable Energy Reviews, vol. 28, pp. 483–493 (2013)
5. Kunte, R.S., Gao, W.: Comparison and review of islanding detection techniques for distributed energy resources. In: 40th North American Power Symposium, pp. 1–8 (2008)
6. Ahmad, K.N.E.K., Selvaraj, J., Abd Rahim, N.: A review of the islanding detection methods in grid connected PV inverters. Renew. Sustain. Energy Rev. 21, 756–766 (2013)
7. Reddy, C.R., Reddy, K.H., Reddy, K.V.S.: Recognition of islanding data for multiple distributed generation systems with ROCOF shore up analysis. In: Satapathy, S., Bhateja, V., Das, S. (eds.) Smart Intelligent Computing and Applications. Smart Innovation, Systems and Technologies, vol. 104, pp. 547-558. Springer, Singapore (2019)
8. Song, J., Song, S.J., Oh, S.D., Yoo, Y.: Optimal operational state scheduling of wind turbines for lower battery capacity in renewable power systems in islands. In: International Conference on Renewable Energy Research and Applications, pp. 164–168 (2016)
9. Li, C., Cao, C., Cao, Y., Kuang, Y., Zeng, L., Fang, B.: A review of islanding detection methods for microgrid. Renew. Sustain. Energy Rev. 35, 211–220 (2014)
10. Fan, Y., Li, C.: Analysis on non detection zone of the islanding detection in photovoltaic grid connected power system. In: International Conference on Advanced Power System Automation and Protection, vol. 01, pp. 275-279 (2011)
11. Mi, Z., Wang, F.: Power equations and non detection zone of passive islanding detection and protection method for grid connected photovoltaic generation system. In: Pacific-Asia Conference on Circuits, Communications and Systems, pp. 360- 363 (2009)
12. Anudeep, B., Nayak, P.K.: A passive islanding detection technique for distributed generations. In: 7th International Conference on Power Systems, pp. 732–736 (2017)
13. Guha, B., Haddad, R.J., Kalaani, Y.: A novel passive islanding detection technique for converter based distributed generation systems. In: IEEE Power & Energy Society Innovative Smart Grid Technologies Conference, pp. 1–5 (2015)
14. Zhao, J., Zhang, D., He, J.: A passive islanding detection method based on interharmonic impedance. In: IEEE Conference on Energy Internet and Energy System Integration, pp. 1–6 (2017)
15. Liu, H., Liu, G., Wang, W., Wu, H.: Research on a novel islanding detection technique. In: Proceeding of the 11th World Congress on Intelligent Control and Automation, pp. 5448-5452 (2014)
16. Isa, A.I.M., Mohamad, H., Yasin, Z.M.: Evaluation on non detection zone of passive islanding detection techniques for synchronous distributed generation. In: IEEE Symposium on Computer Applications & Industrial Electronics, pp. 100–104 (2015)
17. De Mango, F., Liserre, M., Dell'Aquila, A.: Overview of anti islanding algorithms for PV systems. PART I: passive methods. In: 12th International Power Electronics and Motion Control Conference, pp. 1878–1883 (2006)
18. Guha, B., Haddad, R.J., Kalaani, Y.: A passive islanding detection approach for inverter based distributed generation using rate of change of frequency analysis. In: South East Conference, pp. 1–6 (2015)

19. Reddy, C.R., Reddy, K.H.: A passive islanding detection method for neutral point clamped multilevel inverter based distributed generation using rate of change of frequency analysis. Int. J. Electr. Comput. Eng. **08**(04), 1967–1976 (2018)
20. Ghadimi, N., Farhadi, P., Hashemi, F., Ghadimi, R.: Detecting the anti islanding protection based on combined changes of active and reactive output powers of distributed generations. In: 3rd International Conference on Computer Research and Development, vol. 3, pp. 285–289 (2011)
21. Reddy, C.R., Reddy, K.H.: Islanding detection method for inverter based distributed generation based on combined changes of ROCOAP and ROCORP. Int. J. Pure Appl. Math. **117**(19), 433–440 (2017)
22. Tzelepis, D., Dysko, A., Booth, C.: Performance of loss of mains detection in multi generator power islands. In: 13th International Conference on Development in Power System Protection, pp. 1–6 (2016)
23. Jang, S.I., Kim, K.H.: An islanding detection method for distributed generations using voltage unbalance and total harmonic distortion of current. IEEE Trans. Power Deliv. **19**(2), 745–752 (2004)
24. Singam, B., Hui, L.Y.: Assessing SMS and PJD schemes of anti islanding with varying quality factor. In: IEEE International Power and Energy Conference, pp. 196–201 (2006)
25. Khichar, S., Lalwani, M.: An analytical survey of the islanding detection techniques of distributed generation systems. Technol. Econ. Smart Grids Sustain. Energy **03**, 10 (2018)
26. Danandeh, A., Seyedi, H., Babaei, E.: Islanding detection using combined algorithm based on rate of change of reactive power and current THD techniques. In: Asia Pacific Power and Energy Engineering Conference, pp. 1–4 (2012)
27. Jun, L., Xue-liang, H., Xiao-hu, C., Miao, X., Wen, X.: Two islanding detection circuits based on the impedance variation for the micro grid. In: The 2nd International Symposium on Power Electronics for Distributed Generation Systems, pp. 859–863 (2010)
28. Salman, S. K., King, D.J., Weller, G.: New loss of mains detection algorithm for embedded generation using rate of change of voltage and changes in power factors. In: Seventh International Conference on Developments in Power System Protection, pp. 82–85 (2001)
29. Nale, R., Biswal, M.: Comparative assessment of passive islanding detection techniques for microgrid. In: International Conference on Innovations in Information, Embedded and Communication Systems, pp. 1–5 (2017)
30. Raza, S., Mokhlis, H., Arof, H., Mohamad, H., Laghari, J.A.: Prioritization of different passive parameters for islanding detection on the basis of response analysis. In: IEEE International Conference on Power and Energy, pp. 615–619 (2016)
31. Raza, S., Mokhlis, H., Arof, H., Laghari, J.A., Mohamad, H.: A sensitivity analysis of different power system parameters on islanding detection. IEEE Trans. Sustain. Energy **07**(02), 461–470 (2016)
32. Stumpf, P., Nagy, I., Vajk, I.: Novel approach of microgrid control. In: International Conference on Renewable Energy Research and Application, pp. 859–864 (2014)
33. Lin, Z., et al.: Application of wide area measurement system to islanding detection of bulk power systems. IEEE Trans. Power Syst. **28**(02), 2006–2015 (2013)
34. Kumar, K.M., Naresh, M., Singh, N.K., Singh, A.K.: A passive islanding detection approach for distributed generation using rate of change of negative sequence voltage and current. In: IEEE Uttar Pradesh Section International Conference on Electrical, Computer and Electronics Engineering, pp. 356–360 (2016)
35. Chandak, S., Dhar, S., Barik, S.K.: Islanding disclosure for grid interactive PV-VSC system using negative sequence voltage. In: IEEE Power, Communication and Information Technology Conference, pp. 497–504 (2015)
36. Suresh, K., et al.: A passive islanding detection method for hybrid distributed generation system under balanced islanding. Indonesian J. Electr. Eng. Comput. Sci. **14**(1), 9–19 (2019)

37. Ch, R.R., Reddy, K.H.: An efficient passive islanding detection method for integrated DG System with Zero NDZ. Int. J. Renew. Energy Res. **8**(4), 1994–2002 (2018)
38. Rostami, A., Jalilian, A., Naderi, S.B., Negnevitsky, M., Davari, P., Blaabjerg, F.: A novel passive islanding detection scheme for distributed generations based on rate of change of positive sequence component of voltage and current. In: Australasian Universities Power Engineering Conference, pp. 1–5 (2017)
39. Rostami, A., Bagheri, M., Naderi, S.B., Negnevitsky, M., Jalilian, A., Blaabjerg, F.: A novel islanding detection scheme for synchronous distributed generation using rate of change of exciter voltage over reactive power at DG-Side. In: Australasian Universities Power Engineering Conference, pp. 1–5 (2017)
40. Harrouz, A., Colak, I., Kayisli, K.: Control of a small wind turbine system application. In: International Conference on Renewable Energy Research and Applications, pp. 1128–1133 (2016)
41. Mercado, K.D., Jiménez, J., Quintero, M.C.G.: Hybrid renewable energy system based on intelligent optimization techniques. In: International Conference on Renewable Energy Research and Applications, pp. 661–666 (2016)
42. Lin, X., Dong, X., Lu, Y.: Application of intelligent algorithm in island detection of distributed generation. In: Transmission and Distribution Conference and Exposition, pp. 1–7 (2010)
43. Sun, R., Wu, Z., Centeno, V.: Power system islanding detection and identification strategy using topology approach and decision tree. In: IEEE Power & Energy Society General Meeting, pp. 1–6 (2011)
44. Lidula, N.W.A., Perera, N., Rajapakse, A.D.: Investigation of a fast islanding detection methodology using transient signals. In: IEEE Power Energy Society General Meeting, pp. 1–6 (2009)
45. Samantaray, S.R., Arroudi, K., Joos, G., Kamwa, I.: A fuzzy rule based approach for islanding detection in distributed generation. IEEE Trans. Power Deliv. **25**(3), 1427–1433 (2010)
46. Marchesan, G., Muraro, M.R., Cardoso, G., Mariotto, L., De Morais, A.P.: Passive method for distributed generation islanding detection based on oscillation frequency. IEEE Trans. Power Deliv. **31**(01), 138–146 (2016)
47. Reddy, C.R., Reddy, K.H.: Islanding detection for inverter based distributed generation with low frequency current harmonic injection through Q controller and ROCOF analysis. J. Electr. Syst. **14**(02), 179–191 (2018)
48. Mishra, P.P., Bhende, C.N.: A passive islanding detection technique with reduced complexity for distributed generations. In: 7th International Conference on Power Systems, pp. 830–835 (2017)
49. Reddy, S.V.R., Premila, T.R., Reddy, C.R., Gulzar, M.M., Khalid, M.: A new variational mode decomposition-based passive islanding detection strategy for hybrid distributed renewable generations. Arab. J. Sci. Eng. **48**(11), 15435–15443 (2023)
50. Reddy, S.V.R., Premila, T.R., Reddy, C.R., Alharbi, M.A., Alamri, B.: Passive island detection method based on sequence impedance component and load-shedding implementation. Energies **16**(16), 5880 (2023)
51. Reddy, S.V.R., Premila, T.R., Reddy, C.R., Reddy, B.N.: Zero power mismatch islanding detection algorithm for hybrid distributed generating system. Trans. Energy Syst. Eng. Appl. **4**(2), 1–12 (2023)

Rapid Convergence of New FP Iterative Algorithm

Naveen Kumar$^{(\boxtimes)}$ ⓘ and Surjeet Singh Chauhan ⓘ

Division of Mathematics, University Institute of Sciences, Chandigarh University,
Gharuan, Mohali 140413, Punjab, India
imnaveenphd@gmail.com

Abstract. In this paper, a new iterative process called NIP is introduced and some convergence theorems for approximation of fixed points (FPs) of contractive and non-expansive maps are proved. It is also shown that NIP converges rapidly than many existing iterative processes. This new iterative scheme requires least number of iterations as compared with the existing iteration procedures like Picard, Mann, Ishikawa, Noor, Agarwal, Abbas and many others. Further, the same is validated numerically as well as graphically by considering some standard functions.

Keywords: Contractive Mappings · Nonexpansive Mappings · Convex Metric Space · Iterative Processes · Rate of Convergence · Fixed Point

1 Introduction

Iterative schemes are useful tools for the approximation of FPs of nonlinear equations. While studying an iterative process, two criteria are to be followed, one is speed and the other is accuracy. In this direction, various results are proved by researchers like Mann [1], Ishikawa [2,3], Picard (E. Picard 1890), Noor [4], Agarwal et al. [5], Abbas and Nazir [6]. In this research paper, we consider N as the collection of natural numbers, S as a nonvoid closed and convex subset of a real norm space D and $F_H = \{\tau \in S : H\tau = \tau\}$ contains all fixed points of H. Remind that, a point $\tau \in S$ is a FP of $H : S \to S$ if $H\tau = \tau$. The operator H is called the iteration function. A mapping $H : S \to S$ is a contraction if there is $\mu \in (0,1]$ with $\|Hp - Hq\| \leq \mu\|p - q\|$ and nonexpansive if $\|Hp - Hq\| \leq \|p - q\|$ for all $p, q \in S$.

Very few iterative algorithms were proposed for the FP approximation of different classes of operators, some of them are used in this paper to approximate fixed points. A sequence $\{\sigma_n\}$ defined below,

$$\begin{cases} \sigma_1 \in S \\ \sigma_{n+1} = H\sigma_n \end{cases} \quad (n \in N) \tag{1}$$

is called Picard iterative process or the sequence of successive approximation (E. Picard, 1890).

ⓒ The Author(s), under exclusive license to Springer Nature Switzerland AG 2024
S. L. Gundebommu et al. (Eds.): REGS 2023, CCIS 2081, pp. 89–106, 2024.
https://doi.org/10.1007/978-3-031-58607-1_7

If H is the operator defined as,

$$H(\sigma) = \sigma - \frac{f(\sigma)}{f'(\sigma)} \tag{2}$$

where f' is the differentiation of f and σ_0 is the starting point, then Eq. (1) reduces to

$$\sigma_{n+1} = H(\sigma_n) = \sigma_n - \frac{f(\sigma_n)}{f'(\sigma_n)}; n = 0, 1, 2, \ldots \tag{3}$$

which is popularly called Newton's method for solving nonlinear problems. Mann [1] in 1953 devised an iteration process as

$$\begin{cases} \sigma_1 \in S & (n \in N) \\ \sigma_{n+1} = (1 - \omega_n)\, \sigma_n + \omega_n H \sigma_n \end{cases} \tag{4}$$

where $\{\omega_n\}$ be the sequence in $(0, 1)$.

Krasnoselskii [7], in 1955 showed that the Picard iteration Eq. (1) fails to reach to its FP for a nonexpansive map H though it has a unique FP whereas the Mann algorithm Eq. (4) reaches strongly to the FP of H if $\omega_n = \frac{1}{2}$ for $n \in N$. Ishikawa [2,3] in 1974 defined the iteration process as follow,

$$\begin{cases} \sigma_1 \in S \\ \sigma_{n+1} = (1 - \omega_n)\, \sigma_n + \omega_n H \sigma_n & (n \in N) \\ \nu_n = (1 - \delta_n)\, \sigma_n + \delta_n H \sigma_n \end{cases} \tag{5}$$

where $\{\omega_n\}$ and $\{\delta_n\}$ are the sequences in $(0, 1)$.

The iterative procedure Eq. (5) reduces to Mann iterative process Eq. (4) if $\delta_n = 0$, for each $n \in N$.

Many authors like Abbas and Nazir [6], Agarwal et al. [5], Krasnoselskii [7], Noor [4] etc. have studied Mann [1] and Ishikawa [2,3] iterations to approximate FPs of nonexpansive mapping.
Noor [4] defined the iteration procedure in the year 2000 by,

$$\begin{cases} \sigma_1 \in S \\ \sigma_{n+1} = (1 - \omega_n)\, \sigma_n + \omega_n H \sigma_n & (n \in N) \\ \nu_n = (1 - \delta_n)\, \sigma_n + \delta_n H \sigma_n \\ \rho_n = (1 - \epsilon_n)\, \sigma_n + \epsilon_n H \sigma_n \end{cases} \tag{6}$$

where $\{\omega_n\}, \{\delta_n\}$ and $\{\epsilon_n\}$ are the real sequences in $(0, 1)$ and demonstrated that iteration Eq. (6) converges rapidly than Mann [1], Picard (E. Picard, 1890) and Ishikawa [2,3] procedures for contractive mappings. If $\epsilon_n = 0$, then Noor iteration Eq. (6) becomes Ishikawa iteration Eq. (5) and when $\epsilon_n = \delta_n = 0$, then Noor iteration Eq. (6) reduces to the Mann iteration Eq. (6).

Agarwal et al. [5] gave the following iteration in 2007 as,

$$\begin{cases} \sigma_1 \in S \\ \sigma_{n+1} = (1 - \omega_n)H\sigma_n + \omega_n H\nu_n \quad (n \in N) \\ \nu_n = (1 - \delta_n)\,\sigma_n + \delta_n H\sigma_n \end{cases} \tag{7}$$

where the sequences $\{\omega_n\}$ and $\{\delta_n\}$ are in (0, 1) and revealed that this iteration algorithm moving faster than Mann iteration for contractive mappings but at the same rate as that of Picard.

Abbas and Nazir [6] defined the following algorithm in 2014 as,

$$\begin{cases} \sigma_1 \in S \\ \sigma_{n+1} = (1 - \omega_n)H\nu_n + \omega_n H\rho_n \quad (n \in N) \\ \nu_n = (1 - \delta_n)H\sigma_n + \delta_n H\rho_n \\ \rho_n = (1 - \epsilon_n)\,\sigma_n + \epsilon_n H\sigma_n \end{cases} \tag{8}$$

where $\{\omega_n\}, \{\delta_n\}$ and $\{\epsilon_n\}$ are the real sequences in (0, 1) and demonstrated that the iterative procedure Eq. (8) converges more rapidly than Agarwal et al. [5].

Thakur et al. [8] introduced a modified iteration as,

$$\begin{cases} \sigma_1 \in S \\ \sigma_{n+1} = (1 - \omega_n)H\rho_n + \omega_n H\nu_n \quad (n \in N) \\ \nu_n = (1 - \delta_n)\,\rho_n + \delta_n H\rho_n \\ \rho_n = (1 - \epsilon_n)\,\sigma_n + \epsilon_n H\sigma_n \end{cases} \tag{9}$$

where $\{\omega_n\}, \{\delta_n\}$ and $\{\epsilon_n\}$ are the sequences taken in (0, 1).

Motivated from above fixed point iterative processes, we establish a new FP iterative procedure called NIP for finding the FPs of nonexpansive maps having smaller range compared to previous iterative schemes for the sequences as,

$$\begin{cases} \sigma_1 \in S \\ \sigma_{n+1} = H^n\nu_n \quad (n \in N) \\ \nu_n = (1 - \delta_n)H\rho_n + \delta_n H^n\rho_n \\ \rho_n = (1 - \epsilon_n)\,\sigma_n + \epsilon_n H^n\sigma_n \end{cases} \tag{10}$$

where, $\{\delta_n\}$ and $\{\epsilon_n\}$ are the sequences chosen in $\left(\dfrac{1}{2}, 1\right)$.

Our objective is to discuss some convergence theorems with iterative algorithm (10). It is also reveled that NIP converges at a rate rapid than Mann [1], Picard (E. Picard 1890), Ishikawa [2,3], Noor [4], Agarwal et al. [5], Abbas and Nazir [6], Thakur et al. [8] procedures for contractive maps. Moreover, the convergence behavior of all above mentioned iterative processes with NIP is shown in tabular form as well as graphically for a standard function.

2   Preliminaries and Definitions

Let $S_D = \{\sigma \in D : \|\sigma\| = 1\}$ be a unit sphere on a Banach space D. For all $\alpha \in (0, 1)$ if $\|\alpha\sigma_1 + (1 - \alpha)\sigma_2\| < 1$ for $\sigma_1, \sigma_2 \in S_D$ and $\sigma_1 \neq \sigma_2$, then D is known as strictly convex. Moreover, if D is a strictly convex Banach space with $\|\sigma_1\| = \|\sigma_2\| = \|\alpha\sigma_1 + (1-\alpha)\sigma_2\|$ for $\sigma_1, \sigma_2 \in D$ and $\alpha \in (0,1)$, then $\sigma_1 = \sigma_2$. A mapping $H : S \to D$ is said to be demiclosed at $\sigma_2 \in D$ if for every sequence $\{\sigma_n\}$ in S and every $\sigma_1 \in D, \sigma_n \to \sigma_1$ and $H\sigma_n \to \sigma_2$ yields that $\sigma_1 \in S$ and $H\sigma_1 = \sigma_2$.

Definition 2.1. [9] If the sequences $\{r_n\}$ and $\{s_n\}$ converging to r and s respectively with

$$\lim_{n\to\infty} \left| \frac{r_n - r}{s_n - s} \right| = \omega$$

Then, we say the sequence $\{r_n\}$ converges better than $\{s_n\}$ for $\omega = 0$ and the sequences $\{r_n\}$ and $\{s_n\}$ have the same convergence rate for $0 < \omega < \infty$.

Definition 2.2. [10,11] If $H : S \to S$ be a mapping, define

$$\|H^n \sigma_n - \tau\| \le \left(\frac{\omega^n}{1-\omega} \right) \|\sigma_n - \tau\|$$

for all n = 1, 2, 3, . . . and ω is called a Lipschitzian constant.

Definition 2.3. [9] Let the error estimates for two fixed point iterative sequences $\{\sigma_n\}$ and $\{\rho_n\}$, approaching to the common FP τ, be specified as:

$$\|\sigma_n - \tau\| \le r_n \text{ and } \|\rho_n - \tau\| \le s_n$$

for all n = 1, 2, 3, . . . with $\{r_n\}$ and $\{s_n\}$ approaching to 0. If $\{r_n\}$ moves faster than $\{s_n\}$, then $\{\sigma_n\}$ also tending faster than $\{\rho_n\}$ to fixed point τ.

Definition 2.4. [12] Let $\{\sigma_n\}$ be any sequence in Banach space D, then D satisfies the Opial's condition if $\sigma_n \to \sigma_1$ implies that

$$\lim_{n\to\infty} sup \|\sigma_n - \sigma_1\| < \lim_{n\to\infty} sup \|\sigma_n - \sigma_2\| \text{ for all } \sigma_2 \in D \text{ with } \sigma_2 \ne \sigma_2$$

Let us now state some lemmas.

Lemma 2.1. [13] Let H be a nonexpansive mapping on S where $S \ne \phi$ be closed and convex subset of the uniformly convex Banach space D, then $I - H$ is demiclosed at 0.

Lemma 2.2. [14] Let D be a uniformly convex Banach space and $0 < r \leq \alpha_n \leq s < 1$ for all naturals. Let $\{\sigma_n\}$ and $\{\rho_n\}$ are the sequences of D such that $\lim_{n\to\infty} sup \|\sigma_n\| < p$, $\lim_{n\to\infty} sup \|\rho_n\| \leq p$ and , $\lim_{n\to\infty} sup \|\alpha_n\sigma_n + (1 - \alpha_n)\rho_n\| = p$ hold for some positive number p, then $\lim_{n\to\infty} \|\sigma_n - \rho_n\| = 0$.

Lemma 2.3. [5] Consider a Banach space D that satisfies Opial's condition [12]. Let $S \neq \phi$ be closed and convex subset of D and $H : S \to D$ be a map such that $I - H$ is demiclosed at 0. Let $\{\sigma_n\}$ be a sequence in S such that $\lim_{n\to\infty} \|H\sigma_n - \tau\| = 0$. and $\lim_{n\to\infty} \|\sigma_n - \tau\| = 0$ for all $\tau \in F(H)$, then $\{\sigma_n\}$ converges weakly to the fixed point of H.

Lemma 2.4. Let $\{\sigma_n\}$ be a non-negative sequence for which $n_0 \in N$ be such that for all $n \geq n_0$, the inequality $\sigma_{n+1} \leq (1 - \epsilon_n)\sigma_n + \epsilon_n\rho_n$ holds, where $\epsilon_n \in (0,1)$, $\sum_{n=0}^{\infty} \epsilon_n = \infty$ and $\rho_n \geq 0$ for all $n \in N$, then $\lim_{n\to\infty} sup \sigma_n \leq \lim_{n\to\infty} sup \rho_n$.

Lemma 2.5. If ρ be a number satisfying $0 \leq \rho < 1$ and let $\{\varepsilon_n\}$ be a sequence such that $\lim_{n\to\infty} \varepsilon_n = 0$ and $\sigma_{n+1} \leq \rho \sigma_n + \varepsilon_n, n \in N$, then $\lim_{n\to\infty} \sigma_n = 0$.

Lemma 2.6. Let S be a subset of Banach space D and H be a nonexpansive map on S and $\{\sigma_n\}$ be a sequence defined in (10), then $\lim_{n\to\infty} \|H\sigma_n - \sigma_n\| = 0$.

3   Main Results

Theorem 3.1. Let $S \neq \phi$ be closed and convex subset of D. Consider a contraction map H with a contraction factor $\mu \in (0,1)$ and fixed point τ. Let $\{u_n\}$ and $\{\sigma_n\}$ are the sequences defined by the iteration processes (9) and (10) respectively, where $\{\delta_n\}$ and $\{\varepsilon_n\}$ are the sequences chosen in $(\frac{1}{2}, 1)$ for all naturals. Then $\{\sigma_n\}$ converges faster than $\{u_n\}$, that is NIP Eq. (10) converges more rapidly than the modified iterative Algorithm (9).

Proof. From the Theorem 3.1 of Thakur et al. [8], we have:

$$\|u_n - \tau\| \leq \mu[1 - (1 - \mu)\varepsilon_n]^n \|u_1 - \tau\| \text{ for all natural numbers n.}$$

Let $\alpha_n = \mu^n[1 - (1 - \mu)\varepsilon_n]^n \|u_1 - \tau\|$

Let $\{\sigma_n\}$ be a sequence in the iteration process (10), then we obtain

$$\begin{aligned}
\|\rho_n - \tau\| &= \|(1 - \varepsilon_n)\sigma_n + \varepsilon_n H^n\sigma_n - \tau\| \\
&\leq (1 - \varepsilon_n)\|\sigma_n - \tau\| + \varepsilon_n\|H^n\sigma_n - \tau\| \\
&\leq (1 - \varepsilon_n)\|\sigma_n - \tau\| + \varepsilon_n\frac{\omega^n}{1 - \omega}\|\sigma_n - \tau\| \\
&\leq [1 - \varepsilon_n + \varepsilon_n\frac{\omega^n}{1 - \omega}]\|\sigma_n - \tau\| \\
&\leq [1 - \varepsilon_n(1 - \frac{\omega^n}{1 - \omega})]\|\sigma_n - \tau\|
\end{aligned}$$

and

$$\begin{aligned}
\|\nu_n - \tau\| &= \|(1-\delta_n)H\rho_n + \delta_n H^n \rho_n - \tau\| \\
&\leq (1-\delta_n)\|H\rho_n - \tau\| + \delta_n\|H^n\rho_n - \tau\| \\
&\leq (1-\delta_n)\mu\|\rho_n - \tau\| + \delta_n\frac{\omega^n}{1-\omega}\|\rho_n - \tau\| \\
&\leq [(1-\delta_n)\mu + \frac{\omega^n}{1-\omega}\delta_n][1 - \varepsilon_n(1 - \frac{\omega^n}{1-\omega})]\|\sigma_n - \tau\| \\
&\leq \{\mu(1-\delta_n)[1 - \varepsilon_n(1 - \frac{\omega^n}{1-\omega})] + \frac{\omega^n}{1-\omega}\varepsilon_n(1-\delta_n) + \varepsilon_n\delta_n(\frac{\omega^n}{1-\omega})^2\}\|\sigma_n - \tau\|
\end{aligned}$$

Thus

$$\begin{aligned}
\|\sigma_{n+1} - \tau\| &= \frac{\omega^n}{1-\omega}\|\nu_n - \tau\| \\
&= \frac{\omega^n}{1-\omega}\{\mu(1-\delta_n)[1-\varepsilon_n(1-\frac{\omega^n}{1-\omega})] + \frac{\omega^n}{1-\omega}\varepsilon_n(1-\delta_n) + \varepsilon_n\delta_n(\frac{\omega^n}{1-\omega})^2\}\|\sigma_n - \tau\| \\
&= \{\mu(1-\delta_n)\frac{\omega^n}{1-\omega}[1-\varepsilon_n(1-\frac{\omega^n}{1-\omega})] + (\frac{\omega^n}{1-\omega})^2\varepsilon_n(1-\delta_n) + \varepsilon_n\delta_n(\frac{\omega^n}{1-\omega})^2\}\|\sigma_n - \tau\|
\end{aligned}$$

Let

$$\beta_n = \{\mu(1-\delta_n)\frac{\omega^n}{1-\omega}[1-\varepsilon_n(1-\frac{\omega^n}{1-\omega})] + (\frac{\omega^n}{1-\omega})^2\varepsilon_n(1-\delta_n) + \varepsilon_n\delta_n(\frac{\omega^n}{1-\omega})^2\}\|\sigma_1 - \tau\|$$

Then

$$\left|\frac{\beta_n}{\alpha_n}\right| = \left|\frac{\{\mu(1-\delta_n)\frac{\omega^n}{1-\omega}[1-\varepsilon_n(1-\frac{\omega^n}{1-\omega})] + (\frac{\omega^n}{1-\omega})^2\varepsilon_n(1-\delta_n) + \varepsilon_n\delta_n(\frac{\omega^n}{1-\omega})^2\}\|\sigma_1-\tau\|}{\mu^n[1-(1-\mu)\varepsilon_n]^n\|u_1-\tau\|}\right| \to 0 \qquad \text{as}$$

$n \to \infty$.

That is, $\left|\frac{NIP Scheme}{Modified Iteration}\right| \to 0$ whenever n is large.

Hence, the sequence $\{\sigma_n\}$ converges more rapidly than the sequence $\{u_n\}$. Consequently, NIP (10) has better convergence speed than the Modified iterative process (9).

In the following examples, we show that NIP (10) converges at a rate better than many iterations like modified iteration (9), Abbas iteration (8), Agarwal iteration (7), Noor iteration (6), Picard iteration (1), Ishikawa iteration (5) and Mann iteration (4).

Example 3.1. Construct a map $H : [0,1] \to [0,1]$ *with* $H(\sigma) = \frac{e^\sigma}{7}10$. Choose $\delta_n = \varepsilon_n = 0.75$ with $\sigma_1 = 30$. Clearly, H is a contraction with $\mu \in [0.17, 1)$ having a unique fixed point $\tau = 0.1118325$ correct to 7 decimal places.

The following comparative table and figure prove that NIP (10) converges better than the iteration processes defined in (1)–(9) (Table 1).

Table 1. Comparison of NIP with Existing Algorithms

Step	Mann	Ishikawa	Picard	Noor	Agarwal	Abbas	Mod. Iter	NIP
1	0.9900000	0.9900000	0.9900000	0.9900000	0.9900000	0.9900000	0.9900000	0.9900000
2	0.4493425	0.3650461	0.2691234	0.3555435	0.1848269	0.1476213	0.1335658	0.1169677
3	0.2298817	0.1828030	0.1308816	0.1786876	0.1160385	0.1129022	0.1122409	0.1118331
4	0.1518542	0.1316049	0.1139832	0.1301230	0.1120683	0.1118641	0.1118401	**0.1118325**
5	0.1252628	0.1173325	0.1120733	0.1168329	0.1118457	0.1118334	0.1118327	0.1118325
6	0.1163241	0.1133618	0.1118594	0.1131993	0.1118332	**0.1118325**	**0.1118325**	0.1118325
7	0.1133330	0.1122577	0.1118355	0.1122061	0.1118326	0.1118325	0.1118325	0.1118325
8	0.1123336	0.1119507	0.1118328	0.1119346	**0.1118325**	0.1118325	0.1118325	0.1118325
9	0.1119998	0.1118654	**0.1118325**	0.1118604	0.1118325	0.1118325	0.1118325	0.1118325
10	0.1118884	0.1118416	0.1118325	0.1118401	0.1118325	0.1118325	0.1118325	0.1118325
11	0.1118512	0.1118350	0.1118325	0.1118346	0.1118325	0.1118325	0.1118325	0.1118325
12	0.1118387	0.1118332	0.1118325	0.1118331	0.1118325	0.1118325	0.1118325	0.1118325
13	0.1118346	0.1118327	0.1118325	0.1118327	0.1118325	0.1118325	0.1118325	0.1118325
14	0.1118332	0.1118326	0.1118325	**0.1118325**	0.1118325	0.1118325	0.1118325	0.1118325
15	0.1118327	**0.1118325**	0.1118325	0.1118325	0.1118325	0.1118325	0.1118325	0.1118325
16	**0.1118325**	0.1118325	0.1118325	0.1118325	0.1118325	0.1118325	0.1118325	0.1118325

Clearly all the iterative processes Mann [1], Ishikawa [2,3], Picard (E. Picard 1890), Noor [4], Agarwal et al. [5], Abbas and Nazir [6], Thakur et al. [8] and NIP [7] converges to the common fixed point $\tau = 0.1118325$ corrected to 7 decimal places. This comparison indicates that NIP has better convergence rate as it requires only 4 iterations to converge to its fixed point (Fig. 1).

Example 3.2. Consider D = R, S = [1,60] ⊂ D and H: S → S be a map given as

$$H(\sigma) = \sqrt{\sigma^2 - 9\sigma + 54} \text{ for all } \sigma \in S$$

Consider the initial value $\sigma_1 = 30$ and choose $\delta_n = \varepsilon_n = 0.75$. Following table indicates that the Mann [1], Ishikawa [2,3], Picard (E. Picard 1890), Noor [4], Agarwal et al. [5], Abbas and Nazir [6], Thakur et al. [8] and NIP reaches to $\tau = 6$. The above comparison indicates that NIP has better convergence rate as it requires only 5 no. of iterations than all the mentioned existing schemes (Table 2).

The convergence behaviour of NIP (10) with all these iteration methods can also be illustrated from the table given below along with their number of iterations required for getting the result:

Table 2. Comparision of NIP with Existing Processes.

Step	Mann	Ishikawa	Picard	Noor	Agarwal	Abbas	Mod. Iter	NIP
1	30.0000000	30.0000000	30.0000000	30.0000000	30.0000000	30.0000000	30.0000000	30.0000000
2	27.1150452	25.0119824	26.1533936	23.4891.33	24.0503308	22.6107900	21.3066758	19.7207589
3	24.2907437	20.2547559	22.4191761	17.4668190	18.4372719	15.8281562	13.5889959	7.5447211
4	21.5420343	15.8509087	18.8373796	12.3265857	13.3938203	10.2582064	8.1129739	6.0005595
5	18.8892775	12.0133051	15.4696624	8.7275766	9.3725555	7.0018379	6.2256746	**6.0000000**
6	16.3606498	9.0688620	12.4130372	6.9585711	6.9939357	6.1191542	6.0151302	6.0000000
7	13.9954171	7.2820400	9.8166266	6.3102146	6.1862067	6.0112132	6.0009604	6.0000000
8	11.8475686	6.4668031	7.8750567	6.0979255	6.0283693	6.0010243	6.0000607	6.0000000
9	9.9869851	6.1600652	6.7187058	6.0306808	6.0041338	6.0000933	6.0000038	6.0000000
10	8.4900396	6.0537250	6.2187342	6.0095903	6.0005981	6.00000849	6.0000002	6.0000000
11	7.4083030	6.0179028	6.0583865	6.0029956	6.0000864	6.0000007	**6.0000000**	6.0000000
12	6.7246651	6.0059514	6.0148623	6.0009354	6.0000124	**6.0000000**	6.0000000	6.0000000
13	6.3468134	6.0019768	6.0037328	6.0002921	6.0000018	6.0000000	6.0000000	6.0000000
14	6.1586728	6.0006564	6.0009342	6.0000912	6.0000002	6.0000000	6.0000000	6.0000000
15	6.0708846	6.0002179	6.0002336	6.0000284	**6.0000000**	6.0000000	6.0000000	6.0000000
16	6.0313055	6.0000723	6.0000584	6.0000088	6.0000000	6.0000000	6.0000000	6.0000000
17	6.0137535	6.0000240	6.0000146	6.0000027	6.0000000	6.0000000	6.0000000	6.0000000
18	6.0060282	6.0000079	6.0000036	6.0000008	6.0000000	6.0000000	6.0000000	6.0000000
19	6.0026394	6.0000026	6.0000009	6.0000002	6.0000000	6.0000000	6.0000000	6.0000000
20	6.0011551	6.0000008	6.0000002	**6.0000000**	6.0000000	6.0000000	6.0000000	6.0000000
21	6.0005054	6.0000002	**6.0000000**	6.0000000	6.0000000	6.0000000	6.0000000	6.0000000
22	6.0002211	**6.0000000**	6.0000000	6.0000000	6.0000000	6.0000000	6.0000000	6.0000000
23	6.0000967	6.6.0000000	6.0000000	6.0000000	6.0000000	6.0000000	6.0000000	6.0000000
24	6.0000423	6.0000000	6.0000000	6.0000000	6.0000000	6.0000000	6.0000000	6.0000000
25	6.0000185	6.0000000	6.0000000	6.0000000	6.0000000	6.0000000	6.0000000	6.0000000
26	6.0000081	6.0000000	6.0000000	6.0000000	6.0000000	6.0000000	6.0000000	6.0000000
27	6.0000035	6.0000000	6.0000000	6.0000000	6.0000000	6.0000000	6.0000000	6.0000000
28	6.0000015	6.0000000	6.0000000	6.0000000	6.0000000	6.0000000	6.0000000	6.0000000
29	6.0000006	6.0000000	6.0000000	6.0000000	6.0000000	6.0000000	6.0000000	6.0000000
30	6.0000002	6.0000000	6.0000000	6.00000000	6.0000000	6.0000000	6.0000000	6.0000000
31	6.0000001	6.0000000	6.0000000	6.00000000	6.0000000	6.0000000	6.0000000	6.0000000
32	**6.0000000**	6.0000000	6.0000000	6.00000000	6.0000000	6.0000000	6.0000000	6.0000000

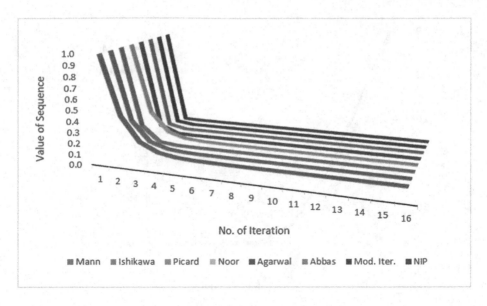

Fig. 1. Comparison of NIP with various existing schemes.

Table 3. Comparison of NIP with Existing Algorithms.

Example-3.1		Example-3.2	
Iteration Procedures	No. of Iterations	Iteration Procedures	No. of Iterations
Mann iteration (M-Iter.)	16	Mann iteration (M-Iter.)	32
Ishikawa iteration (I-Iter.)	15	Mann iteration (M-Iter.)	22
Picard iteration (P-Iter.)	09	Picard iteration (P-Iter.)	21
Noor iteration (N-Iter.)	14	Noor iteration (N-Iter.)	20
Agarwal iteration (Ag-Iter.)	08	Agarwal iteration (Ag-Iter.)	15
Abbas iteration (Abb-Iter.)	06	Abbas iteration (Abb-Iter.)	12
Modified iteration (Mod-Iter.)	06	Modified iteration (Mod-Iter.)	11
NIP	04	NIP	05

Hence NIP has better rate of convergence than the other existing procedures (see Table 3).

Now, various cases are shown in the following graphs which show that NIP (10) converges more rapidly than various processes like Mann [1], Ishikawa [2,3], Picard (1), Noor [4], Agarwal et al. [5], Abbas and Nazir [6] and Thakur et al. [8] respectively (Figs. 2, 3, 4, 5, 6, 7 and 8).

1. It is clear from the Theorem 3.1 that NIP (10) convergence more rapidly than Thakur et al. [8] iteration process i.e. $|\frac{NIP\ Scheme}{Modified\ Iter.}| \to 0$ for large n.

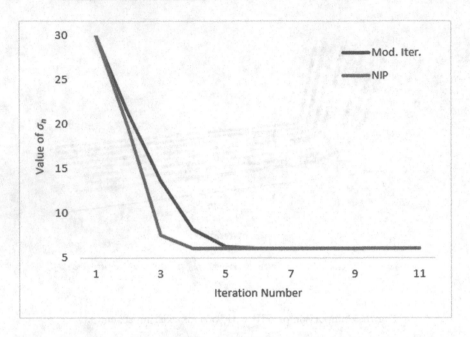

Fig. 2. Graphical Representation of Convergence Behaviour.

In a similar way, for large n, the comparisons of all the iteration procedures (1)-(9) with the NIP (10) can be plotted separately as shown in the following graphs:

2. $|\frac{NIP \ Scheme}{Abbas \ et \ al.}| \to 0$, so NIP (10) converges speedier than Abbas and Nazir [6] iteration.

3. $|\frac{NIP \ Scheme}{Agarwal \ Iter.}| \to 0$, so NIP (10) converges faster than Agarwal et al. [5] iteration.

4. $|\frac{NIP \ Scheme}{Noor \ Iter.}| \to 0$, thus NIP (10) convergence faster than Noor [4] iteration.

5. $|\frac{NIP \ Scheme}{Picard \ Iter.}| \to 0$, so NIP (10) has better convergence rate than Picard iteration (E. Picard, 1890).

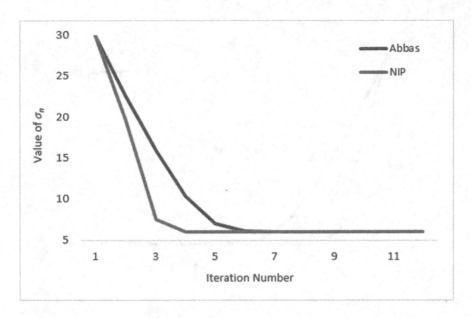

Fig. 3. Graphical Representation of Convergence Behaviour.

6. $\left|\dfrac{NIP\ Scheme}{Ishikawa\ Iter.}\right| \to 0$, thus NIP (10) converges with better speed than Ishikawa [2,3] iteration.

7. $\left|\dfrac{NIP\ Scheme}{Mann\ Iter.}\right| \to 0$, so NIP (10) converges at a rate better than Mann iteration [1].

From the above graphs, it is observed that NIP (10) converges rapidly than all the iterative schemes (1)-(9) (see Fig. 9).
Following combined graph clearly describes the convergence behavior of NIP among the iterations i.e. Mann [1], Ishikawa [2,3], Picard (E. Picard 1890), Noor [4], Agarwal et al. [5], Abbas and Nazir [6] and Thakur et al. [8] iteration processes:

We now establish some convergence theorems using NIP (10).

Theorem 3.2. Let $H : D \to D$ be a map on a norm linear space D with a fixed point τ satisfying $\|H\nu - \tau\| \le \mu \|\nu - \tau\|$, where $\mu \in [0,1)$. Let $\{\sigma_n\}$ be a sequence defined in NIP (10) where $\{\delta_n\}$ and $\{\varepsilon_n\} \in (\frac{1}{2},1)$ and $\sum_{n=0}^{\infty} \delta_n = \infty$, $\sum_{n=0}^{\infty} \varepsilon_n = \infty, \sum_{n=0}^{\infty} \delta_n\varepsilon_n = \infty$, then NIP (10) converges to the fixed point of H.

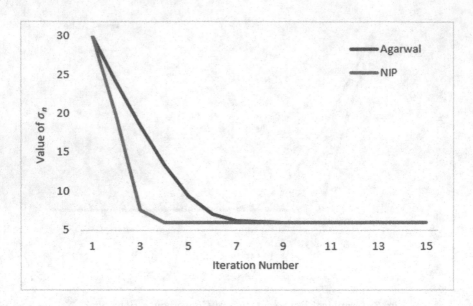

Fig. 4. Graphical Representation of Convergence Behaviour.

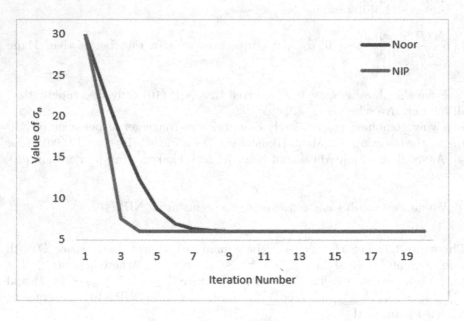

Fig. 5. Graphical Representation of Convergence Behaviour.

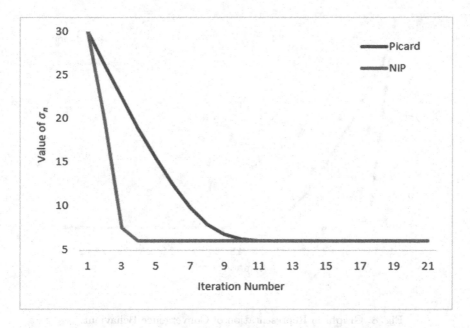

Fig. 6. Graphical Representation of Convergence Behaviour.

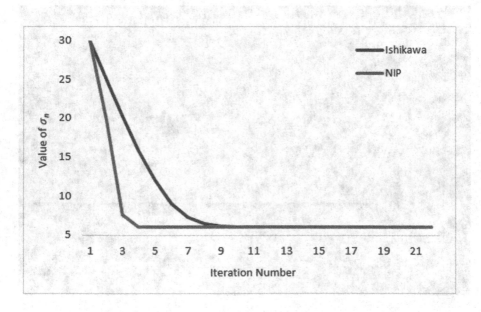

Fig. 7. Graphical Representation of Convergence Behaviour.

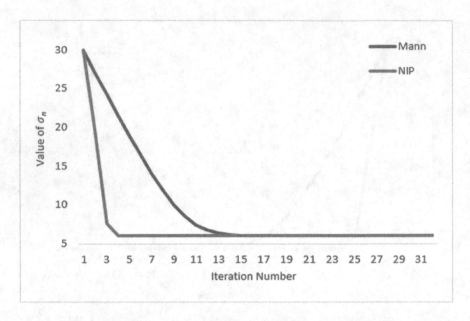

Fig. 8. Graphical Representation of Convergence Behaviour.

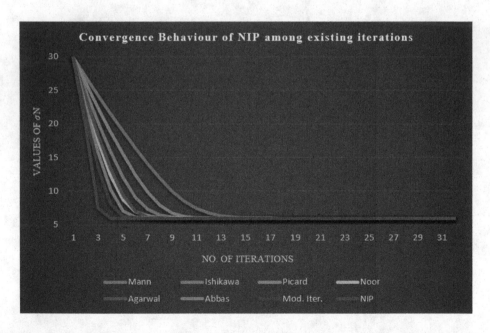

Fig. 9. Combined Graph of NIP's Convergence Behavior among Existing Iterations.

Proof. Let $\{\sigma_n\}$ be a sequence defined in (10), then we have

$$
\begin{aligned}
\|\rho_n - \tau\| &= \|(1 - \varepsilon_n)\sigma_n + \varepsilon_n H^n \sigma_n - \tau\| \\
&\leq (1 - \varepsilon_n)\|\sigma_n - \tau\| + \varepsilon_n\|H^n \sigma_n - \tau\| \\
&\leq (1 - \varepsilon_n)\|\sigma_n - \tau\| + \varepsilon_n \frac{\omega^n}{1 - \omega}\|\sigma_n - \tau\|, \text{ by Definition 2.2} \\
&\leq (1 - \varepsilon_n)\|\sigma_n - \tau\| + \varepsilon_n L\|\sigma_n - \tau\|, \text{ where } L = \frac{\omega^n}{1 - \omega} \\
&= (1 - (1 - L)\varepsilon_n)\|\sigma_n - \tau\|
\end{aligned}
\tag{11}
$$

and

$$
\begin{aligned}
\|\nu_n - \tau\| &= \|(1 - \delta_n)H\rho_n + \delta_n H^n \rho_n - \tau\| \\
&\leq (1 - \delta_n)\|H\rho_n - \tau\| + \delta_n\|H^n \rho_n - \tau\| \\
&\leq (1 - \delta_n)\mu\|\rho_n - \tau\| + \delta_n \frac{\omega^n}{1 - \omega}\|\rho_n - \tau\|, \text{ by Definition 2.2} \\
&\leq ((1 - \delta_n)\mu + L\delta_n)(1 - (1 - L)\varepsilon_n)\|\sigma_n - \tau\|, \text{ where } L = \frac{\omega^n}{1 - \omega}
\end{aligned}
\tag{12}
$$

Also, using (11) and (12), we derive

$$
\begin{aligned}
\|\sigma_{n+1} - \tau\| &= -\frac{\omega^n}{1 - \omega}\|\nu_n - \tau\| \\
&= L\|\nu_n - \tau\|, \text{ where } L = \frac{\omega^n}{1 - \omega} \\
&= ((1 - \delta_n)\mu + L\delta_n)(1 - (1 - L)\varepsilon_n)\|\sigma_n - \tau\| \\
&= [\mu L(1 - \delta_n)(1 - L\delta_n) + L^2 \delta_n(1 - \varepsilon_n) + L^3 \delta_n \varepsilon_n]\|\sigma_n - \tau\|
\end{aligned}
$$

Since $\sum_{n=0}^{\infty} \delta_n = \infty, \sum_{n=0}^{\infty} \varepsilon_n = \infty, \sum_{n=0}^{\infty} \delta_n \varepsilon_n = \infty$, then by Lemma 2.4, we have

$$
\lim_{n \to \infty} \|\sigma_n - \tau\| = 0.
$$

Hence the result.

Theorem 3.3. Let S be a non-void, closed and convex subset of a norm linear space D and let H be a nonexpansive map on S. Let $\{\sigma_n\}$ be a sequence in NIP (10) and $F(H) \neq \phi$, then $\lim_{n \to \infty} \|\sigma_n - \tau\| = 0$ exists for all $\tau \in F(H)$ for all naturals n with replacing H^n by H in NIP (10).

Proof. For arbitrarily chosen $\sigma_1 \in S$, NIP is defined as:

$$
\begin{aligned}
\sigma_{n+1} &= H^n \nu_n \\
\nu_n &= (1 - \delta_n)H\rho_n + \delta_n H^n \rho_n \\
\rho_n &= (1 - \varepsilon_n)\sigma_n + \varepsilon_n H^n \sigma_n
\end{aligned}
\tag{13}
$$

where $\{\delta_n\}$ and $\{\varepsilon_n\}$ are the sequences chosen in $(\frac{1}{2}, 1)$.
Replacing H^n by H in iterative process (13), we obtain

$$\sigma_{n+1} = H\nu_n$$
$$\nu_n = H\rho_n$$
$$\rho_n = (1 - \varepsilon_n)\sigma_n + \varepsilon_n H\sigma_n \qquad (14)$$

where the sequence $\{\varepsilon_n\}_{n\epsilon N} \epsilon (\frac{1}{2}, 1)$.
From (14), we have

$$
\begin{aligned}
\|\rho_n - \tau\| &= \|(1 - \varepsilon_n)\sigma_n + \varepsilon_n H\sigma_n - \tau\| \\
&\leq (1 - \varepsilon_n)\|\sigma_n - \tau\| + \varepsilon_n\|H\sigma_n - \tau\| \\
&\leq (1 - \varepsilon_n)\|\sigma_n - \tau\| + \varepsilon_n\|\sigma_n - \tau\| \text{ as H is nonexpansive} \\
&= \|\sigma_n - \tau\| \qquad (15)
\end{aligned}
$$

and

$$
\begin{aligned}
\|\nu_n - \tau\| &= \|H\rho_n - \tau\| \\
&= \|\rho_n - \tau\| \text{ as H is nonexpansive} \\
&= \|\sigma_n - \tau\| \qquad (16)
\end{aligned}
$$

Thus, using (15) and (16), we derive
$$\|\sigma_{n+1} - \tau\| = \|H\nu_n - \tau\| \leq \|\sigma_n - \tau\| \text{ as H is nonexpansive}$$
Thus, $\lim_{n\to\infty} \|\sigma_n - \tau\|$ exists for all $\tau \epsilon F(H)$.

Theorem 3.4. Let $(D, \|.\|)$ be a real norm space and $H : D \to D$ be a map satisfying $\|H\nu - \tau\| \leq \mu\|\nu - \tau\|$, where $\mu \epsilon [0, 1)$. Let $\{\sigma_n\}$ be a sequence defined in (10) where $\{\delta_n\}$ and $\{\varepsilon_n\} \epsilon (\frac{1}{2}, 1)$ and $\sum_{n=0}^{\infty} \delta_n = \infty, \sum_{n=0}^{\infty} \varepsilon_n = \infty, \sum_{n=0}^{\infty} \delta_n\varepsilon_n = \infty$, then the iterative scheme (10) converges to the fixed point of H, that is $\lim_{n\to\infty} \|\sigma_n - \tau\| = 0$.

Proof. Let $\{\sigma_n\}$ be a sequence defined in (12), then

$$
\begin{aligned}
\|\rho_n - \tau\| &= \|(1 - \varepsilon_n)\sigma_n + \varepsilon_n H\sigma_n - \tau\| \\
&\leq (1 - \varepsilon_n)\|\sigma_n - \tau\| + \varepsilon_n\|H\sigma_n - \tau\| \\
&\leq (1 - \varepsilon_n)\|\sigma_n - \tau\| + \varepsilon_n\mu\|\sigma_n - \tau\| \\
&\leq [1 - (1 - \mu)\varepsilon]\|\sigma_n - \tau\| \qquad (17)
\end{aligned}
$$

and using (17), we derive

$$\begin{aligned}
\|\nu_n - \tau\| &= \|H\rho_n - \tau\| \\
&\leq \mu\|\rho_n - \tau\| \\
&\leq \mu[1 - (1 - \mu)\varepsilon_n]\|\sigma_n - \tau\|
\end{aligned} \tag{18}$$

Also, using (18), we have

$$\begin{aligned}
\|\sigma_{n+1} - \tau\| &= \|H\nu_n - \tau\| \\
&\leq \mu\|\nu_n - \tau\| \\
&\leq \mu^2[1 - (1 - \mu)\varepsilon_n]\|\sigma_n - \tau\|
\end{aligned}$$

From Lemma 2.4, we get

$$\lim_{n \to \infty} \|\sigma_n - \tau\| = 0$$

4   Conclusions

In this research, a new iterative process is defined and compared with various existing algorithms. Also, some convergence theorems are proved by using this iterative process. Analytical and graphical approach investigate that the iterative algorithm (10) converges more rapidly than existing iterative schemes like Mann [1], Ishikawa [2,3], Picard (E. Picard, 1890), Agarwal et al. [5], Abbas and Nazir [6], Noor [4] and Thakur et al. [8] iteration procedures.

References

1. Mann, W.R.: Mean value methods in iteration. Proc. Am. Math. Soc. 4, 506–510 (1953)
2. Ishikawa, S.: Fixed points by a new iteration method. Proc. Am. Math. Soc. 44, 147–150 (1974)
3. Ishikawa, S.: Fixed points and iteration of a nonexpansive mapping in a Banach space. Proc. Am. Math. Soc. 59(1), 65–71 (1976)
4. Noor, M.A.: New approximation schemes for general variational inequalities. J. Math. Anal. Appl. 251(1), 217–229 (2000)
5. Agarwal, R.P., O'Regan, D., Sahu, D.R.: Iterative construction of fixed points of nearly asymptotically nonexpansive mappings. J. Nonlinear Convex Anal. 8(1), 61–79 (2007)
6. Abbas, M., Nazir, T.: A new faster iteration process applied to constrained minimization and feasibility problems. Mat. Vesn. 66(2), 223–234 (2014)
7. Krasnoselskii, M.A.: Two observations about the method of successive approximations. (Russian) Uspenski Mathematiki Nauka, 10, 123–127 (1955)
8. Thakur, B.S., Thakur, D., Postolache, M.: A new iteration scheme for approximating fixed points of nonexpansive mappings. Filomat 30(10), 2711–2720 (2016)

9. Berinde, V.: Picard iteration converges faster than Mann iteration for a class of quasicontractive operators. Fixed Point Theory Appl. **2**, 97–105 (2004)
10. Kumar, N., Chauhan (Gonder), S.S.: Analysis of Jungck-Mann and Jungck-Ishikawa Iteration schemes for their speed of convergence, vol. 2050, p. 020011. AIP Publishing (2018)
11. Chauhan (Gonder) S.S., et al.: New fixed point iteration and its rate of convergence. Optimization **72**, 2415–2432 (2022)
12. Opial, Z.: Weak convergence of the sequence of successive approximations for nonexpansive mappings. Bull. Am. Math. Soc. **73**, 591–597 (1967)
13. Goebel, K., Kirk, W.A.: Topics in Metric Fixed Point Theory. Cambridge Studies in Advanced Mathematics 28. Cambridge University Press (1990)
14. Schu, J.: Weak and strong convergence to fixed points of asymptotically nonexpansive mappings. Bull. Aust. Math. Soc. **43**, 153–159 (1991)

A Case Study: Design Based Model of Electric Vehicle

M. Lakshmi Swarupa[1]([✉]), K. Rayudu[2], S. Sunanda[3], G. Divya[1], and M. Rajitha[1]

[1] CVR College of Engineering, Ibrahimpatnam, Hyderabad, India
swarupamalladi@gmail.com, m.rajitha@cvr.ac.in
[2] BVRIT, Narsapur, Hyderabad, India
rayudu.katuri@bvrit.ac.in
[3] St. Martins Engineering College, Hyderabad, India
ssunandaeee@smec.ac.in

Abstract. A design-based model can effectively involve various phases, such as defining Functional Specifications, Design Specifications, Testing and Authentication, and Implementation. However, this paper's case study will only cover the first two stages. The results obtained from this study will provide insights into how system designers can make decisions based on complex execution that may accurately rely both basic and complex designs, approaching practical models with great precision. However, there is a trade-off between precision and design complexity when making decisions. While necessary models with accurateness are considered, determining values, especially in the initial stage of model development, can be challenging. Additionally, simulating these accurate models can be time-consuming. Therefore, a more detailed level of simulation model is necessary. Consequently, system designers require a comprehensive understanding of system power flow during the initial modeling phase. In the second phase, it is crucial to design more accurate models for different systems, including selecting appropriate parameters for energy management systems and types of converters. This paper emphasizes that a model behavior which has high reliability enables necessary adjustments.

Keywords: Electric vehicle · Model Based system · Vehicle Dynamics · Field-oriented control strategy · MATLAB – SIMULINK

1 Introduction

The improvement of electric vehicles includes significantly developed to time to time. It is a vehicle performance analysis done with different parameters. Various vehicle modeling, including multi-physics, steady state, mathematical domain transient, physical modeling, and vehicle dynamics modeling, are available. It has been many changes in the field in present. Given the existence of several architectures, simulation plays a crucial role in finalizing the performance and controls [5]. Vehicle modeling also varies in terms of accuracy and fidelity, ranging from component modeling to system level modeling. The choice of model type and modeling tools depends on the work for vehicle performance.

© The Author(s), under exclusive license to Springer Nature Switzerland AG 2024
S. L. Gundebommu et al. (Eds.): REGS 2023, CCIS 2081, pp. 107–117, 2024.
https://doi.org/10.1007/978-3-031-58607-1_8

When developing an electric vehicle, the developer must make decisions regarding the battery's state of charge and the quantity of batteries. The choice of motor is crucial and necessary. If the performance of the latest EVs, which have replaced the motor, is not matched with that of the vehicles being analyzed, it poses a challenge. Understanding how an EV operates under various driving and environmental conditions is essential to optimize its health, performance, and safety. Parameters such as temperature, road condition, road elevation, and driving pattern must be taken into account to validate the reliability and performance of the vehicle. The model-based design process is employed to enhance the development process of EV conversion. This approach enables engineers to make informed decisions, resulting in cost and time savings [2–4]. Through this process, simulation can be conducted using different specifications and scenarios. The initial step in the model-based EV conversion prototype is EV modeling, which includes the traction, components, and power flow of the EV.

2 Modelling of EV Components

The study in this paper revolves around the Toyota Prius. The primary focus is on the electrical components, namely the Battery, DC/DC Converters, Motor Drive, and Generator Drive. The performance testing of prototype vehicles has shown that they are on par with conventional vehicles in terms of cost and time. Therefore, the analytical modeling of electric vehicles has played a crucial role in determining the future of transportation systems. [5]. The description of each component is done in below (Fig. 1):

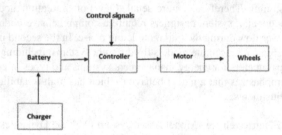

Fig. 1. General diagram of EV

2.1 Battery

The SimPowerSystem library utilizes a simplistic battery model, incorporating a limited number of parameters. Specifically, a 201 V, 6.5 Ah NiMH battery, commonly employed in the Toyota Prius, is taken into account. This battery proves to be more than sufficient in providing accurate precision.

2.2 DC/DC Converter

The Motor and Generator supply the DC bus Voltage, which is then regulated to maintain a constant level. A constant DC source is employed on the DC bus side to keep the voltage fixed. The required current is drawn from the battery, while a filter is utilized to disrupt the algebraic loop.

2.3 Motor Drive

The system is subjected to an electrical torque by the Induction motor. The necessary torque for the system will be determined by the energy management system. The torque regulator assumes that a reference torque is directly applied to the motor shaft. The calculation of the DC bus current can be achieved by measuring both the shaft speed and the DC bus voltage.

3   Induction Motor Related to EVs Modeling

The EV's induction motor operates at a torque lower than its maximum capacity [3]. Numerous advancements have been made in electrical drive systems [6], largely due to the utilization of semiconductors [7]. Field Oriented Control (FOC) has become increasingly crucial in controlling electrical drive controllers, as it relies on the manipulation of current and voltage in three phases. By adjusting the frequency and voltage using the v/f method, the AC machine can be controlled similarly to a DC machine, enabling the attainment of the desired speed and torque [8]. Consequently, AC machines can serve as the primary drive-in electric vehicles (Fig. 2).

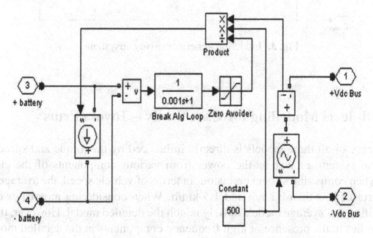

Fig. 2. Induction motor with Vector controlled drive.

3.1   Vehicle Dynamics

Vehicle dynamic control is simplified yet highly valid [9]. Tire characteristics play a crucial role in vehicle dynamics, and reliable tire characteristics and parameters are essential for successful simulation models [10]. Vehicle dynamics encompasses the actions of running, controlling, and terminating the vehicle, which are fundamental functions in automobiles [11]. The performance of a vehicle can be assessed through its dynamics, as the speed of an electric vehicle must maintain the equilibrium between the driving force produced by an induction motor and the kinetic forces exerted on the vehicle needs to be maintained (see Fig. 3).

3.2 Generator Drive

The representation of the Induction Motor model is the same as that of the Induction Generator. The production of Electrical Power involves extracting negative torque from the energy management system. At the beginning, it is assumed that the control system of the generator is flawless, enabling the direct application of the reference torque to the mechanical system. The determination of the corresponding current is then carried out based on the

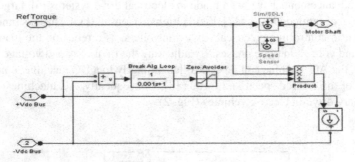

Fig. 3. Induction generator drive subsystem.

4   Multi- level Modelling for Case Study – Toyota Prius

The accuracy of all three models is directly influenced by the torque and speed of the mechanical system, as well as the power from various components of the electrical system, when comparing system precision. In terms of vehicle speed, the average model has an error of less than 2 km/h and 1.5 km/h. When considering motor power, both the simplified and average models closely match the detailed model. However, the main difference lies in the presence of high frequency components in the detailed model due to the The inverter's switching frequency has been altered. The simplified and detailed models exhibit a maximum error rate of less than 10%, with an average error rate of merely 5%.

4.1   Vehicle Torque of Simple, Average and Detailed Model Comparison

The torque discrepancy for each of the three models is a mere 5%. The results indicate that the simplified model showcases a more pronounced reaction to the required torque specified by the energy management system. Conversely, the average model exhibits a steady increment in torque towards the target value with remarkable precision when compared to the desired model. Furthermore, the detailed model stands out due to the high-frequency signal generated by the electrical system's switching frequency. This research paper focuses on examining the simplified, average, and detailed analyses in relation to the Electric vehicle in MTALAB – SIMULINK.

4.2 Basic

In the Basic type, with real-time data not taken into account. Parameters of car speed versus time are considered.

4.3 Mean Values

The closed loop system takes into account the road parameters to determine speed error select an appropriate control strategy, either power-split or torque-split. The average-values algorithm is then utilized to minimize the speed error.

4.4 Detailed Analysis

In this particular category, the closed loop system (torque-split control strategy) takes into account the on-road parameters. To minimize speed error, the average-values algorithm is implemented (Fig. 4).

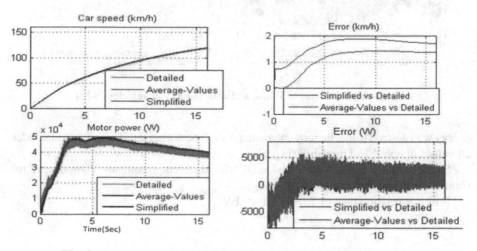

Fig. 4. Power versus time(secs) for simple, average and detailed models

Analysis System simulation of EV enables the achievement of optimal performance for the energy management system, mechanical system, and other components. The minimal simulation time in this model facilitates the adjustment of the system management of

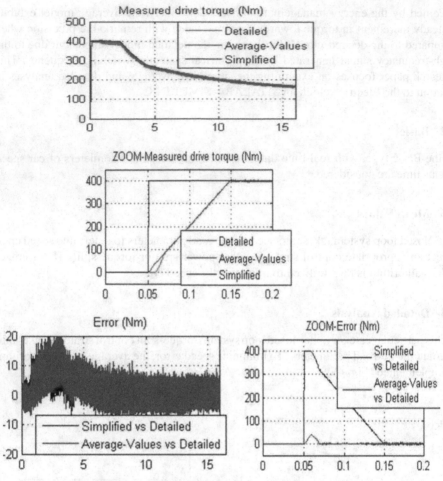

Fig. 5. View in terms of Torque and error in Electric Vehicles (N-m)

energy w.r.t desired value, thereby enhancing overall performance. This analysis includes all values of uphill and downhill (Fig. 5).

Signals of the battery are considered with different frequency components for simulation study (Fig. 6).

Baseline performance of Toyota Prius - EV:

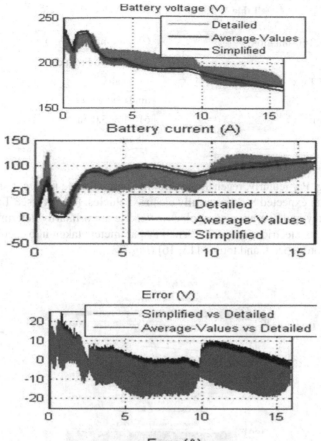

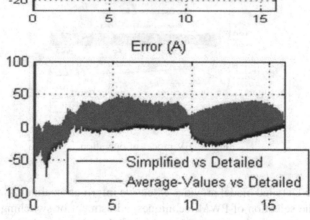

Fig. 6. Error of speed, torque values with different models

4.5 Regenerating Braking Mode

Significance EVs and HEVs have grown as the demand for fuel-efficient and environmentally friendly vehicles continues to rise. These vehicles will contribute to reducing costs. Despite the disappointing mileage, urban driving compensates for the energy required

Table 1. Ratings of Electric Vehicle

Engine - Type of Transmission	CV automatic (speed)
Type of Drive	Wheel with Front drive
Capacity of Fuel Tank	11.3 gal
Type of Fuel	unleaded Regular type
Range (Cty/Hwy)	610.2/565.0 mi
Size of Engine	1.8 l

through braking. Particularly, regenerative braking will enhance fuel economy by 30% in hybrids and an expected 8–26% in fully electric vehicles. [12–14] (see Table 1).

The efficiency and operation of an electric vehicle are primarily determined by the performance of its electric motor drive. The key parameters taken into account include stator current, rotor speed, and torque [15, 16] (Fig. 7).

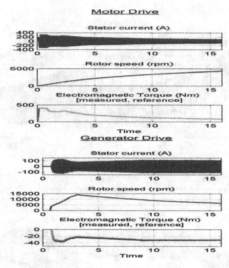

Fig. 7. EV modes of operation

Analysis: The Detailed model provides extensive information about power converters. It allows for the selection of PWM techniques, adjustment of switching frequency for DC/DC Converters, tuning of hysteresis bands and the selection of values aid in achieving reliability in circuits, addressing issues with various disabilities and disturbances. Although these values, it requires a significant amount of time (Fig. 8).

Analysis: Estimated table, various values are to facilitate comprehension. Based on the results, it can be inferred that system modeling does not exhibit any flickering. The models speculated the mean value system modeling is comparatively superior (see Table 2).

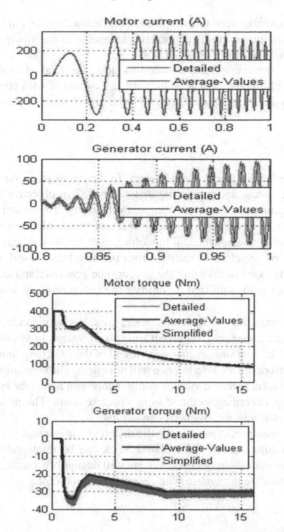

Fig. 8. Generator parameters like Torque, power, current and speed versus time(secs)

Table 2. Comparison Tables for different systems

System	Est Time	Mech. Dyna	DC bus Volt	M/G stator currents	P_L	THD
Basic	11	Ok	NO	NO	NO	NO
Mean	56 (Acc) 256 (Basic)	Ok	Ok	Ok	Ok	NO
Modified	1440 (acce)	Ok	Ok	Ok	Ok	Ok

(i) A disparity exists between the simplified and detailed models in terms of the DC machine. Primarily, the estimated model is incapable of accurately representing the current of either the Motor or Generator.

(ii) In the case of the average and detailed models, the detailed model showcases a component with frequency. The values for the current remain consistent between the average and detailed models, although the phases may yary due to mechanical speed.

5 Conclusion

Engineers will design and develop the system. In the initial stage, engineers focus on understanding the system and simulating its workings. This objective is crucial for the system for management of energy to function effectively. The model in simplified is used for simulation of system with power, torque and speed values. Since the estimated time is minimal with the mathematical modeling, multiple systems can easily analyzed and the observations are close to reality. From the simplified model, obtained results are with case study assist in selecting the appropriate generator and motor. The model with average values is then utilized to validate the behavior of the energy and system management.

Subsequently, engineers utilizes the model to select semiconductor components based on average and instantaneous current and voltage values. Loss evaluation is necessary for designing the heat sink. Proper adjustment of the switching frequency is crucial to ensure that electromagnetic interference will not effect. The simulation of the detailed model allows engineers to test of various components with adjust the system with management of energy, according to the system's requirements. The mean time for high accuracy, which interfaces with outperforms setups.

Therefore, the isolation of models for the different components of drive is essential for initial system adjustments. The isolated block can be reintegrated. Finally, after completing the simulation, the obtained system will be highly accurate. The next step is to experimentally realize the system with reduced time and cost.

References

1. Vinot, E., Scordia, J., Trigui, R., Jeanneret, B., Badin, F.: Model simulation, validation and case study of the 2004 THS of Toyota Prius. Int. J. Veh. Syst. Model. Test. 3(3), 139–167 (2008)
2. Himmler, A., Lamberg, K., Beine, M.: Hardware-in-the-loop testing in the context of ISO 26262. SAE Technical Paper 2012–01–0035 (2012). https://doi.org/10.4271/2012-01-0035
3. Larminie, J., Lowry, J.: Electric Vehicle Technology Explained. John Wiley & Sons, UK (2013). ISBN: 0-470-85163-5
4. Patil, K., Muli, M., Zhu, Z.: Model-based development and production implementation of motor drive controller for hybrid electric vehicle. SAE Technical Paper 2013–01–0158 (2013). https://doi.org/10.4271/2013-01-0158
5. Unnewehr, L., Knoop, C.: Electrical component modeling and sizing for EV simulation. SAE Technical Paper 780215. https://doi.org/10.4271/780215
6. Bose, B.K.: Modern Power Electronics and AC Drives, vol. 123. Prentice Hall, Upper Saddle River (2002)

7. Markel, T., et al.: A systems analysis tool for advanced vehicle modeling. J. Power **110**(2), 255–266 (2002)
8. Hadboul, R.M., Ali, A.M.: Modeling, simulation and analysis of electric vehicle driven by induction motor. In: IOP Conference Series: Materials Science and Engineering (2021). https://doi.org/10.1088/1757-899X/1105/1/012022
9. Ammon, D.: Modellbildung und Systementwicklung in der Fahrzeugdynamik. Stuttgart: B. G. Teubner (1997)
10. A. Lutz a, J. Rauh b & W. Reinalter c a Robert Bosch GmbH, Abstatt, Germany b Daim- ler AG, Sindelfingen, Germany c MAGNA STEYR, Graz," Vehicle System Dynamics", Austria Published online (2008). https://doi.org/10.1080/00423110801925393"
11. Kutluay, E., Winner, H.: Validation of vehicle dynamics simulation models – a review. Veh. Syst. Dyn. **52**(2), 186–200 (2014). https://doi.org/10.1080/00423114.2013.868500
12. Vignesh, R., Benin, S.R.: Design and analysis of regenerative braking system of all-terrain vehicle 1R, vol. 7, Issue 06 (2020). ISSN- 2394–5125
13. Swarupa, M.L.: Performance evaluation and energy management system for parallel hybrid electric vehicle. In: 2023 7th International Conference on Green Energy and Applications (ICGEA), Singapore, pp. 224–229 (2023). https://doi.org/10.1109/ICGEA57077.2023.101 25733
14. Latha, K., Swarupa, M.L.: Design and implementation of power conditioning for distribution network V2G to electric vehicle and DC charging system (2020)
15. Divya, G., Venkata, P.S.: Design and modeling of hybrid electric vehicle powered by solar and fuel cell energy with quadratic buck/boost converter. WSEAS Trans. Circ. Syst. **22**, 41–54 (2023). https://doi.org/10.37394/23201.2023.22.7
16. Sreeshobha, E., Lakshmi, G.S.: Performance analysis and comparison of P.I. controller and ANN controller of bidirectional DC/DC converter for hybrid electricvehicle system. In: 2022 International Conference on Breakthrough in Heuristics And Reciprocation of Advanced Technologies (BHARAT), Visakhapatnam, India, pp. 31–36 (2022). https://doi.org/10.1109/BHARAT53139.2022.9932632

Power Electronics and Renewable
Energy Technologies

Asymmetrical Current Source Multilevel Inverter with Multicarrier PWM Strategies

N. Muruganandham[1]($\boxtimes$) (iD) and T. Suresh Padmanabhan[2] (iD)

[1] Department of Electrical Engineering, Anna University, Chennai, Tamil Nadu, India
nveeramurugan@gmail.com
[2] Department of EEE, E.G.S. Pillay Engineering College, Nagapattinam, Tamil Nadu, India

Abstract. Because of lower stress in terms of rate of voltage, current, and harmonic content, most Inverters which include Multilevel are frequently used in applications like different power converters. This study primarily focuses on a new parallel H-bridge Current Source Multilevel Inverter (CSMLI) circuit arrangement for power system applications. The suggested circuit operates by coupling a DC source to the H-bridge CSI to produce the multi-level output current waveform. Othered power device count, inverter losses, and other novel features can be found in the suggested circuit. The effectiveness of the selected nine-level H-bridge CSI is evaluated through the MATLAB/Simulink program and a Multicarrier PWM control approach.

Keywords: Multicarrier PWM · Harmonics · Current Source Inverter · H-bridge · Total Harmonic Distortion

1 Introduction

Conversion of Direct Current (DC) to Alternating Current (AC) is essential in today's electric power system. Multilevel inverters have been an important piece of technology for a variety of applications over the past decade [1–6]. This includes variable frequency drives, wind, marine and renewable energy sources. The widespread adoption of multi-level inverters (MLIs) has led to their status as a well-established technology. The output voltage of a standard converter can swing between +Vdc and Vdc. The large size of the filters is required to obtain the desired sinusoidal waveform from the two-level output voltage. Also, high-rated components in low-level inverter cause higher switching losses and it is unsuitable for use in high-voltage/power applications [7–10]. The multilayer inverters use several low dc voltage sources to operate the power electronic switches in a certain order, resulting in a stepped output voltage waveform. Multilevel inverters can be categorized as either symmetric or asymmetric. H-bridge DC voltage source values are equivalent in symmetric multilevel inverter topologies; however, they are not in asymmetric topologies. Asymmetric multilevel inverters are capable of generating a broader spectrum of output voltages while utilizing an equal or reduced quantity of power electronic switches. In a CHB asymmetrical multi-level inverter, the determination of the magnitude of DC power sources can be accomplished using two distinct methods:

© The Author(s), under exclusive license to Springer Nature Switzerland AG 2024
S. L. Gundebommu et al. (Eds.): REGS 2023, CCIS 2081, pp. 121–135, 2024.
https://doi.org/10.1007/978-3-031-58607-1_9

binary or trinary. More levels can be generated by a ternary setup than by a binary one. When determining the size and cost of an MLI, the most important aspects are the use of simple control techniques, the use of few semiconductor switches, the use of DC voltage sources, and the use of driving circuits. Recent years have seen various presentations of topologies for symmetric and asymmetric architecture that make use of less number switches [11–17].

2 Current Source Multilevel Inverter

2.1 General

A Direct Current (DC) source supplies electrical power to a Current Source Inverter (CSI). In the context of adjustable speed drives (ASDs), it is common practice to employ an AC/DC rectifier featuring a substantial inductor to serve as the DC source. This configuration serves the purpose of averting the development of a steady current supply. A current source inverter (CSI) commonly has a voltage-boosting capability, enabling its output voltage to surpass the DC-link voltage at its greatest value. The essential circuit layout of an H-bridge Current Source Inverter (CSI) is illustrated in Fig. 1. The inverter under consideration generates a current waveform with three distinct levels, which can be likened to the levels denoted as +I, 0, and −I depicted in Table 1.

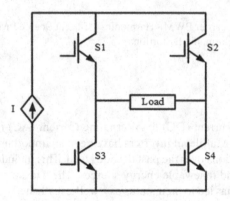

Fig. 1. Basic circuit of H-bridge CSI

Table 1. Switching of H-bridge basic CSI

S1	S2	S3	S4	Output
ON	OFF	OFF	ON	+I
ON	OFF	ON	OFF	OFF
OFF	ON	OFF	ON	OFF
OFF	ON	ON	OFF	−I

2.2   Proposed Current Source Multilevel Inverter Topology

The architecture of the chosen parallel connected Current Source nine-level Inverter is illustrated in Fig. 2. The depicted image illustrates a circuit model that incorporates the coupling of an H-bridge Current-Source Inverter (CSI), unidirectional regulated power devices, and a direct current (DC) supply. The DC module runs at intermediate levels to generate waveforms with nine levels of output. The chosen parallel connected CSI is equipped with parallel Direct Current (DC) sources, each having an amplitude of one-fourth (1/4) of the total current (I). In a previous era, all direct current sources were interconnected. Consequently, the circuit no longer necessitates the utilization of isolated Direct Current (DC) current sources [18, 19]. The operational principle of the suggested CSI topology is presented in Table 2.

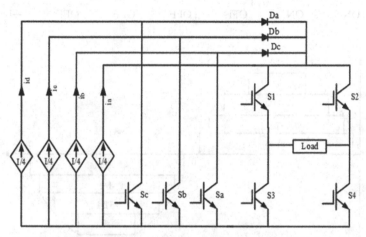

Fig. 2. Proposed parallel connected Current Source nine-level H-bridge Inverter.

Table 2 illustrates switching sequence for the generation of a nine-level waveform through the utilization of parallel connected Current Source nine-level H-bridge Inverter. The switching operation for levels +I, +3I/4, +I/2, +I/4, 0, −I/4, −I/2, −3I/4, and −I are denoted by their respective switching sequences performed using Multicarrier Pulse Width Modulation (PWM) methods.

During the H-bridges positive level, switches S1 and S4 are in the ON state, resulting in a current flow of +I through the load. In a similar manner, when considering the −I level, the H-bridge switches S2 and S3 are activated, resulting in the flow of current through the load in the negative direction, denoted as −I. At the +3I/4 level, the H-bridge switches S1 and S4 are activated, along with switch Sc, to counteract a portion of the current flow equivalent to I/4. Consequently, the resulting current passing through the load is +3I/4. Similarly, to counteract a portion of the current −I/4 at the −3I/4 level, the switch Sc is activated to neutralize it. Additionally, the H-bridge switches S2 and S3 are activated, resulting in a current flow of −3I/4 through the load.

During the specified duration at half level, the H-bridge switches S1 and S4 are activated, while the circuit switches Sb and Sc are turned on to counteract the current

Table 2. Switching Pattern of proposed parallel connected CSI

S1	S2	S3	S4	S_a	S_b	S_c	Output
ON	OFF	OFF	ON	OFF	OFF	OFF	I
ON	OFF	OFF	ON	OFF	OFF	ON	3I/4
ON	OFF	OFF	ON	OFF	ON	ON	I/2
ON	OFF	OFF	ON	ON	ON	ON	I/4
OFF	OFF	ON	ON	ON	ON	ON	0
OFF	ON	ON	OFF	ON	ON	ON	−I/4
OFF	ON	ON	OFF	OFF	ON	ON	−I/2
OFF	ON	ON	OFF	OFF	OFF	ON	−3I/4
OFF	ON	ON	OFF	OFF	OFF	OFF	−I

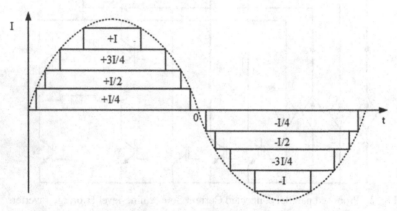

Fig. 3. Nine-level output model of a proposed Current Source Inverter

flow of I/2 through the diodes Db and Dc. The present electrical current passing through the load is equal to negative one-half of the total current, denoted as −I/2. The generation of the inverted level −I/2 is achieved by activating H-bridge switches S2 and S3, and by neutralizing the current flow −I/2 through the circuit switches Sb and Sc. As a result, the load current is −I/2. During the I/4 level, it is observed that the H-bridge switches S1 and S4 are activated, while the circuit switches Sa, Sb, and Sc are also turned on to counteract the maximum current flow. The expression "3I/2" can be interpreted as a mathematical equation. The present magnitude of electric current passing through the load is equal to one-fourth of the total current, denoted as I. Similarly, the generation of the inverted level −I/4 involves the activation of H-bridge switches S2 and S3, as well as the circuit switches Sa, Sb, and Sc to neutralize the maximum current flow of −3I/2. Consequently, the resulting load current is −I/4.

At the zero level, the H-bridge operates by either activating the upper switches S1 and S2, or the lower switches S3 or S4, to effectively cancel out the current flow and bring it to a neutral state. The zero level is determined by the activation of the switch located on the bottom switches S3 and S4. Figure 3 illustrates the nine-level output model of a Current Source Inverter.

3 Hybrid Multicarrier PWM Strategies

The proposed architecture involves the selection of multiple carrier-based Pulse Width Modulation (MCPWM) algorithms, which are chosen based on different carrier signals. Carrier pulse-width modulation (CPWM) approaches employ many carrier waveforms, including saw-tooth and triangle waves. A wide range of control options exists for carrier signals, encompassing parameters such as amplitude, carrier offset, frequency, and phase modulation for each carrier. The Multicarrier Pulse Width Modulation (MCPWM) method evaluates carrier signals that have been vertically displaced in relation to a reference waveform. To transmit the voltage or current at the output of an m-level inverter, the Multicarrier Pulse Width Modulation (MCPWM) method makes use of M − 1 triangular carriers. The proposed architecture utilizes a total of eight triangular carriers in a nine-level configuration. The amplitudes, denoted as Ac, and frequencies, denoted as fc, of the carrier waveforms are equivalent. The reference waveforms are characterized by their amplitude Aref and frequency Fre. Whenever there is a requirement for a switching sequence to be implemented for the output of the inverter, the response of the comparison is decoded. Equations (1) and (2) are employed for the computation of the frequency modulation index (mf) and amplitude modulation index (m$_a$) [20–22].

$$m_f = \frac{f_c}{f_{ref}} \tag{1}$$

$$m_a = \frac{A_{ref}}{(m-1)A_c} \tag{2}$$

In the proposed hybrid multicarrier PWM, Phase Disposition (PD), Phase Opposite Disposition (POD), Alternative Phase Disposition (APOD), Carrier Overlapping, and Variable frequency strategies are used. The reference waveform is sinusoidal. The output response of the proposed CSI is shown for ma = 0.9, mf = 40, and m = 9 for all multicarrier strategies.

3.1 Phase Disposition (PD)

In Phase Disposition (PD), the triangle carriers located above and below the zero reference are in phase. The following pictures illustrate the implementation of the Phase Disposition approach using a sinusoidal reference signal. The figures depict the voltage waveform and its accompanying harmonic spectrum, along with an analysis of the current. Figure 4(a) illustrates the responses of the proportional-derivative (PD) controller with a sine signal.

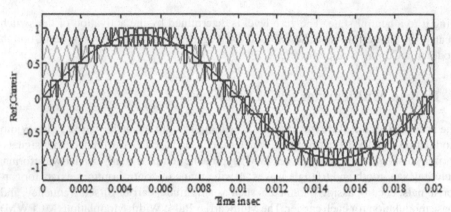

Fig. 4. Gate Pulse Generation of PDPWM strategy for nine-level H-bridge CSI with sine reference

3.2 Phase Opposition Disposition (POD)

Phase Opposition Disposition (POD) is when the carriers with 180° phase angle with sinusoidal reference at low level. Figure 5(a) displays POD waveform using a sine reference.

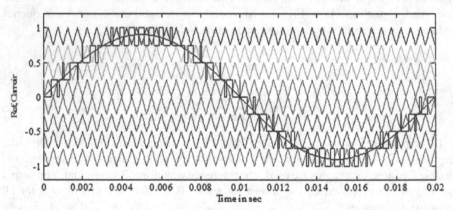

Fig. 5. PODPWM strategy PWM pulses for multi-level (9) H-bridge Current Source Inverter with sine reference

3.3 Alternative Phase Opposition Disposition (POD)

Every carrier is phase-shifted 180° away from its neighboring carrier in an Alternative Phase Opposition Disposition (APOD). The reactions of APOD with sine reference are shown in Fig. 6(a).

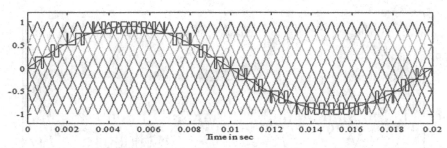

Fig. 6. Gate Pulse Generation of APODPWM strategy for nine-level H-bridge CSI with sine reference

3.4 Carrier Overlapping (CO)

Figure 7(a) displays the vertical offset of carriers for the selected inverter using the COPWM approach and a sine reference. The bands representing the carriers, denoted as Ac, are strategically arranged to achieve overlap in an m-level inverter carrier overlapping pulse width modulation (PWM) scheme. This modulation scheme utilizes m − 1 carriers, each having the same frequency (fc) and comparable peak-to-peak amplitude. Each carrier is spaced vertically apart by an angle of Ac/2. The middle of the carrier signals is where the amplitude of Am and frequency of FM are located.

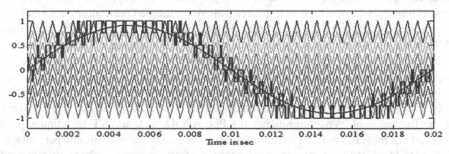

Fig. 7. Gate Pulse Generation of COPWM strategy for nine-level H-bridge CSI with sine reference

3.5 Variable Frequency (VF)

The current ripple could be predicted using a theoretical analysis before the creation of the pulses. Figure 8(a) depicts the VFPWM method of sine references. Compared to intermediary switches in PWM employing constant frequency carriers, the number of switches for the upper and lower devices of the chosen inverter is significantly higher. The number of switches for each switch is balanced by using a variable frequency PWM technique. To achieve balance in the distribution of switches across all switch units, it is necessary to appropriately increase the carrier frequency of the intermediary switches.

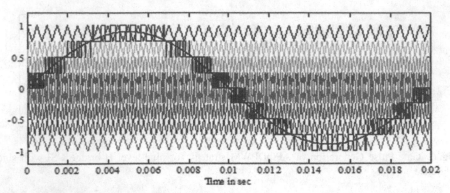

Fig. 8. Gate Pulse Generation of VFPWM strategy for nine-level H-bridge CSI with sine reference

4 Results and Discussion

The performance evaluation of proposed parallel connected Current Source nine-level H-bridge Inverter using multicarrier PWM strategies are depicted in Figs. 9, 10, 11, 12 and 13. Phase voltage, phase voltage and total harmonic disorder responses of PD, POD, APOD, COPWM and VFPWM are show in the figures respectively.

Figure 9(a), (b), and (c) display the analysis of phase voltage, phase current, and frequency spectrum in the PD technique using a sine reference. In the examination of the spectrum, it was observed that the cumulative harmonic contribution amounted to 15.66%. Except for the fundamentals, the relative magnitude of the harmonics is less than 0.5%. Additionally, the highest number of low-order harmonics has been eliminated.

Figures 10(a), (b), and (c) illustrate the analysis of phase voltage, phase current, and frequency spectrum in the POD technique with a sine reference. The frequency spectrum analysis involves the elimination of lower-order harmonics, resulting in a reduced presence of higher-order harmonics. Additionally, all harmonic orders are below 3%, except for the fundamental. When comparing the PD, APOD, CO, and VF techniques, it is observed that the THD is 12.69%, representing the most favorable harmonic percentage.

Figure 11(a), (b), and (c) depict the analysis of phase voltage, phase current, and harmonic spectrum about the APOD method with a sine reference. The lower-order harmonics have been eliminated from the analysis of this spectrum. The utilization of this particular approach yields an aggregate harmonic distortion of 14.35%.

Figures 12(a), (b), and (c) depict the analysis of phase voltage, phase current, and harmonic spectrum in the context of the carrier overlapping approach with a sine reference. The spectrum analysis indicates that the harmonic distortion is 20.17% more when considering both lower and higher-order harmonics compared to the phase disposition technique.

Figure 13(a), (b), and (c) illustrate the utilization of the Variable Frequency method, which involves the study of the phase voltage, phase current, and harmonic spectrum using a sine reference. Please examine the following spectrum analysis: The utilization of the variable Frequency approach resulted in the total elimination of all harmonics of lower order. A total harmonic order of 12.95% was produced.

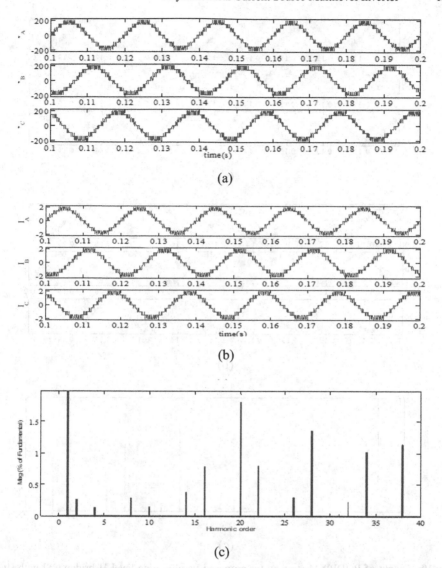

Fig. 9. Output of PDPWM strategy for proposed Parallel nine-level H-bridge CSI with sine reference with ma = 0.9, mf = 40 and m = 9, (a) phase voltage, (b) phase current (c) Harmonic order

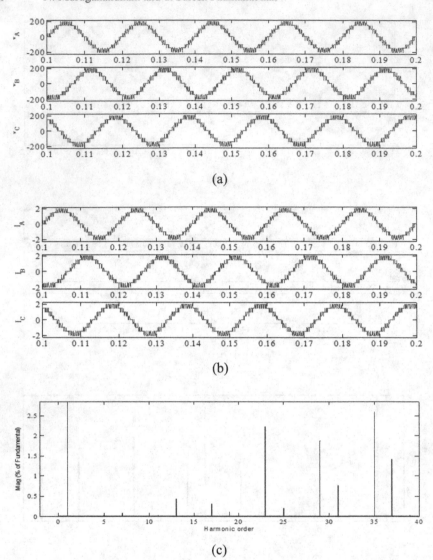

(a)

(b)

(c)

Fig. 10. Output of PODPWM strategy for proposed Parallel nine-level H-bridge CSI with sine reference with ma = 0.9, mf = 40 and m = 9, (a) phase voltage, (b) phase current (c) Harmonic order

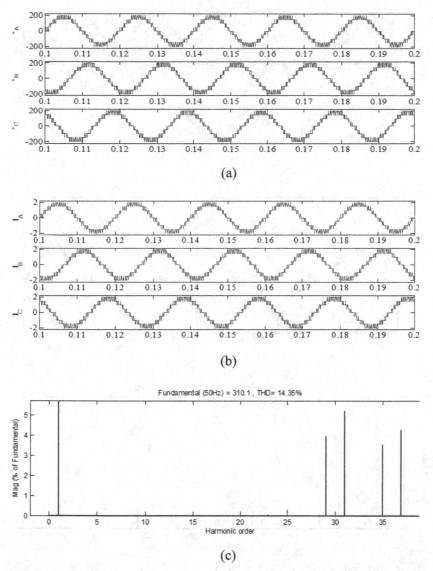

Fig. 11. Output of APODPWM strategy for proposed Parallel nine-level H-bridge CSI with sine reference with ma = 0.9, mf = 40 and m = 9, (a) phase voltage, (b) phase current (c) Harmonic order

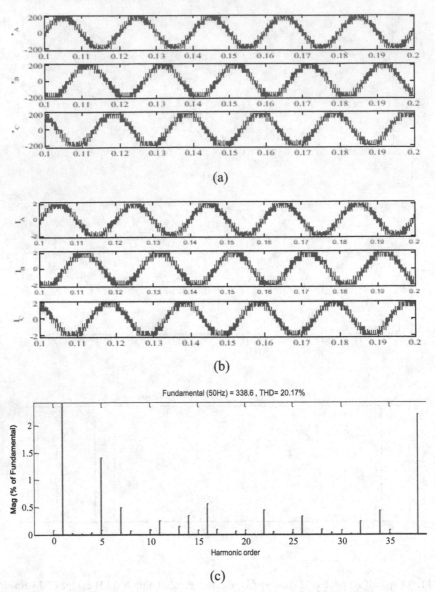

Fig. 12. Output of COPWM strategy for proposed Parallel nine-level H-bridge CSI with sine reference with ma = 0.9, mf = 40 and m = 9, (a) phase voltage, (b) phase current (c) Harmonic order

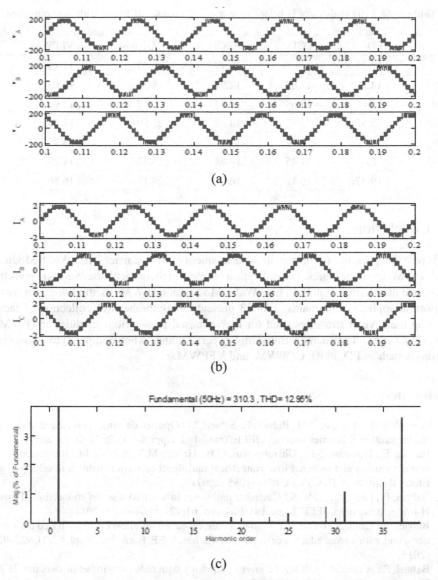

(a)

(b)

(c)

Fig. 13. Output of VFPWM strategy for proposed Parallel nine-level H-bridge CSI with sine reference with ma = 0.9, mf = 40 and m = 9, (a) phase voltage, (b) phase current (c) Harmonic order

Table 3. % THD analysis of H-bridge nine-level CSI using MCPWM with Sine reference

m_a	PD	POD	APOD	COPWM	VFPWM
1	12.21	11	10.19	16.55	11.19
0.95	14.29	12.03	12.94	18.34	12.98
0.9	15.66	12.69	14.35	20.17	12.95
0.85	15.55	13.32	14.45	21.89	12.75
0.8	15.07	13.49	13.97	23.8	13.7
0.75	16.33	14.45	14.48	26.12	14.27
0.7	19.47	16.12	16.85	28.18	16.86

5 Conclusion

This research presents a comprehensive assessment of Multicarrier Pulse Width Modulation (PWM) methodologies. The utilization of relevant theoretical discoveries reveals the existence of a distinct optimal Pulse Width Modulation (PWM) methodology for many important application domains. Table 3 presents a comprehensive collection of facts. Among the several strategies used for carrier-based pulse width modulation (PWM), the APOD method has demonstrated higher performance when compared to alternative methods such as PD, POD, COPWM, and VFPWM.

References

1. Alishah, R.S., Hosseini, S.H., Babaei, E., Sabahi, M.: Optimal design of new cascaded switch-ladder multilevel inverter structure. IEEE Trans. Ind. Appl. **64**(3), 2072–2080 (2017)
2. Babaei, E., Hosseini, S.H., Gharehpetian, G.B., Haque, M.T., Sabahi, M.: Reduction of dc voltage sources and switches in asymmetrical multilevel converters using a novel topology. Electr. Power Syst. Res. **77**(8), 1073–1085 (2007)
3. Babaei, E., Laali, S., Alilu, S.: Cascaded multilevel inverter with series connection of novel H-bridge basic units. IEEE Trans. Ind. Electron. **61**(12), 6664–6671 (2014)
4. Babaei, E., Laali, S., Bayat, Z.: A single-phase cascaded multilevel inverter based on a new basic unit with a reduced number of power switches. IEEE Trans. Ind. Appl. **62**(2), 922–929 (2015)
5. Babaei, E.: A cascade multilevel converter topology with reduced number of switches. IEEE Trans. Power Electron. **23**(6), 2657–2664 (2008)
6. Boost, M.A., Ziogas, P.D.: State-of-the-art carrier PWM techniques: a critical evaluation. IEEE Trans. Ind. Appl. **24**(2), 271–280 (1988)
7. Bowes, S.R.: New sinusoidal pulse width-modulated invertor. In: Proceedings of the Institution of Electrical Engineers, pp. 1279–1285. IET Digital Library (1975)
8. Ceglia, G., Guzmán, V., Sanchez, C., Ibanez, F., Walter, J., Giménez, M.I.: A new simplified multilevel inverter topology for DC–AC conversion. IEEE Trans. Power Electron. **21**(5), 1311–1319 (2006)
9. Choi, W.K., Kang, F.S.: H-bridge based multilevel inverter using PWM switching function. In: INTELEC 2009–31st Proceedings on International Telecommunications Energy, Incheon, Korea, pp. 1–5 (2009)

10. Dixon, J.W., Ortuzar, M., Moran, L.: Drive system for traction applications using 81 level converters. In: Proceedings of IEEE Vehicle Power and Propulsion, Paris, France, pp. 6–8. IEEE (2004)
11. Ebrahimi, J., Babaei, E., Gharehpetian, G.B.: A new multilevel converter topology with reduced number of power electronic components. IEEE Trans. Ind. Electron. **59**(2), 655–667 (2012)
12. Farhadi Kangarlu, M., Babaei, E., Laali, S.: Symmetric multilevel inverter with reduced components based on non-insulated dc voltage sources. IET Power Electron. **5**(5), 571–581 (2012)
13. Hammond, P.W.: A new approach to enhance power quality for medium voltage AC drives. IEEE Trans. Ind. Appl. **33**(1), 202–208 (1997)
14. Hernández, F., Morán, L., Espinoza, J., Dixon, J.: A multilevel active front end rectifier with current harmonic compensation capability. In. 30th Annual Conference on IEEE Industrial Electronics Society, Busan, Korea, pp. 1446–1451. IEEE (2004)
15. Marchesoni, M., Mazzucchelli, M., Tenconi, S.: A nonconventional power converter for plasma stabilization. IEEE Trans. Power Electron. **5**(2), 212–219 (1990)
16. McGrath, B.P., Holmes, D.G.: Multicarrier PWM strategies for multilevel inverters. IEEE Trans. Ind. Electron. **49**(4), 858–867 (2002)
17. Rodriguez, J., Lai, J.S., Peng, F.Z.: Multilevel inverters: a survey of topologies, controls, and applications. IEEE Trans. Ind. Electron. **49**(4), 724–738 (2002)
18. Su, G.J.: Multilevel DC-link inverter. IEEE Trans. Ind. Appl. **41**(3), 848–854 (2005)
19. Suroso, Noguchi, T.: New H-bridge multilevel current-source PWM inverter with reduced switching device count. In: Proceedings of Power Electronics Conference (IPEC), Sapporo, pp.1228–1235 (2010)
20. Suroso, Nugroho, D.T., Noguchi, T.: H-bridge-based five-level current source inverter for grid connected photovoltaic power conditioner. TELKOMNIKA **11**(3), 489–494 (2013)
21. Waltrich, G., Barbi, I.: Three-phase cascaded multilevel inverter using power cells with two inverter legs in series. IEEE Trans. Ind. Appl. **57**(8), 2605–2612 (2010)
22. Zhong, D., Tolbert, L.M., Chiasson, J.N., Ozpineci, B., Li, H., Huang, A.Q.: Hybrid cascaded H bridges multilevel motor drive control for electric vehicles. In: 37th International Proceedings on IEEE Power Electronics Specialists, Jeju, Korea (South) , pp. 1–6. IEEE (2006)

Optimizing the Technological Efficiency of Hybrid Photovoltaic Systems to Fulfill the Energy Requirements of Emergency Shelters for Refugees of the Ukrainian War

Milan Belik[1], Olena Rubanenko[2,3,4], G. Sree Lakshmi[5], and M. Lakshmi Swarupa[5(✉)]

[1] Faculty of Electrical Engineering, University of West Bohemia, Pilsen, Czech Republic
belik4@fel.zcu.cz
[2] Vinnytsia National Technical University, Vinnytsia, Ukraine
olenarubanenko@ukr.net
[3] Igor Sikorski Institute National Technical University in Kyiv, Kyiv, Ukraine
[4] Research and Innovation Centre for EE University of West Bohemia, Pilsen, Czech Republic
[5] CVR College of Engineering, Ibrahimpatnam, Hyderabad, India
g.sreelakshmi@cvr.ac.in, swarupamalladi@gmail.com

Abstract. Hybrid photovoltaic systems have become a common solution for reducing energy consumption in specific objects and for customers in the present time. The efficiency of the entire system also depends on the technology of the battery inverter used. Generally, DC coupled inverters are known to be more energy efficient. However, in certain cases, AC coupled systems can provide better results. The ongoing aggression by Russia against Ukraine has escalated the problem of internal migration, which can only be solved by constructing new communities of emergency shelters. The integration of these units into the overloaded and damaged distribution grids in Ukraine must be carefully planned to limit power consumption and injection. Significant savings can be achieved by properly applying AC or DC coupled systems.

This article discusses this phenomenon based on specific real cases that are defined by consumption profiles, battery storage system management, climate conditions, and PV system design. Simulations presented in the article demonstrate the expected annual energy flows for both technologies in a model situation. The differences between DC coupling and AC coupling solutions are explained through in-depth analyses of inverter behavior, battery behavior, charging strategies, charging losses, discharging losses, state of charge (SOC), cycle load, and the correlation between own consumption and inverter self-consumption. The results show that choosing the right battery inverter technology can lead to significant energy savings from the installed PV system. In certain cases, AC coupled systems not only offer higher flexibility and modularity but also higher energy efficiency for the hybrid system, lower grid feed-in, and better economic profitability.

Keywords: Photovoltaic system · Hybrid inverters · Emergency shelter · AC coupling · DC coupling

© The Author(s), under exclusive license to Springer Nature Switzerland AG 2024
S. L. Gundebommu et al. (Eds.): REGS 2023, CCIS 2081, pp. 136–147, 2024.
https://doi.org/10.1007/978-3-031-58607-1_10

1 Introduction

PV systems are frequently used as systems with hybrid nature to enhance performance. These systems typically involve a PV inverter that is connected to battery storage but does not communicate with the rest of the system [1].

The traditional approach, known as AC coupling, allows for greater flexibility in terms of layout and component selection. However, it also results in higher energy conversions. For instance, common current inverters like the Fronius Symo Hybrid have an efficiency of around 97% when used as a photovoltaic inverter and 96% as a battery inverter [2–5].

On the other hand, inverters are included for different supplies. This is because the inverter and battery are integrated into a single compact unit. In certain cases, factors such as building layout, load-bearing capacity, and safety regulations may prevent the implementation of this theoretically superior solution. It is worth noting that while DC coupling systems have higher conversion efficiency, they may result in lower or less optimal utilization of solar energy [6–9].

Ukraine has led to a region of the country. This has resulted in issues that can only be addressed by rapidly constructing energy-efficient buildings that are simple and cost-effective.

Through various projects, it has been observed while still providing a satisfactory standard of living. However, it is important to note that even as a community unit, these buildings may not achieve complete energy self-sufficiency. Connecting these communities, which consisted of buildings, to the already overloaded Ukrainian distribution networks poses a challenge. These networks were already in a poor state prior to the war and are currently being further damaged by targeted missile attacks. Therefore, achieving a perfect balance between consumption and production is crucial [10].

2 The Implementation of Hybrid Photovoltaic System in Emergency Housing for War Refugees

Both systems present a distinct approach to harnessing photovoltaic energy and employ a unique method for its storage. This disparity arises from variations in losses and their temporal progression. While hybrid PV systems predominantly utilize LiFePo or LiN-iMnCo batteries, gel lead batteries (Moll 3-OPzV) were specifically selected for the cost-effective emergency housing solution catering to war refugees. Another important feature of lead-acid batteries for this application is less sensitivity to extreme temperatures (in this case, especially low) [11–13]. Figure 1 shows the discharge characteristics of the used battery and its cycle life.

To eliminate any variations resulting from the diverse technological features of various manufacturers, Fronius Symo Hybrid inverters were employed in both AC and DC coupling scenarios [14]. These inverters provide all the necessary capabilities. The plots are depicted in Fig. 2.

The chosen inverter has the versatility to function as a dedicated PV inverter, seamlessly integrated including battery that has BMS through different couplings. Additionally, it can operate as a dedicated BMS of inverter when interfaced with solar supply.

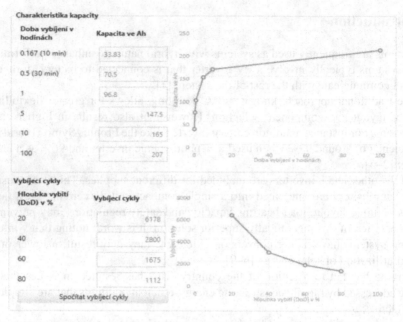

Fig. 1. Discharging characteristic of MOL 3-OPzV (PV*SOL Premium 2023)

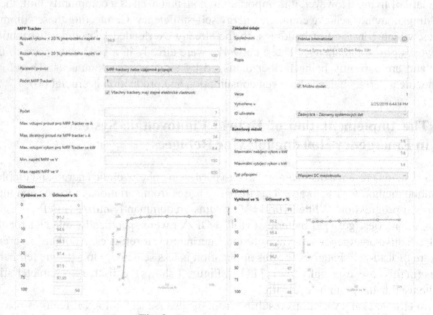

Fig. 2. Inverter plot versus time

Furthermore, it serves like standalone inverter with hybrid nature, effectively combining the functionalities of both photovoltaic and battery inverters within a single unit. The general model of this device is illustrated in Fig. 3.

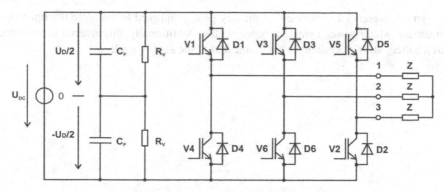

Fig. 3. Topology of 3 phase inverter

The standard inverter can be controlled by adjusting the pulse width in relation to the duration of power being on. This control mechanism allows for easy adjustment of the pulse size to regulate the power supplied with control applicable to both PV inverters and battery inverters. Duty cycle can be set using the basic equation of 1.

$$\text{Duty Cycle} = \frac{pulse\ width}{Period} * 100\% \tag{1}$$

RC is the time constant considered.

$$V_{ripple} = \frac{e^{\frac{-d}{f_{PWM}\ RC}} \cdot (e^{\frac{1}{f_{PWM}\ RC}} - e^{\frac{d}{f_{PWM}\ RC}}) \cdot (1 - e^{\frac{d}{f_{PWM}\ RC}})}{1 - e^{\frac{1}{f_{PWM}\ RC}}} \cdot V_+ \tag{2}$$

Duty cycle varies from 0 to 1.

$$fc = \frac{1}{2\pi\sqrt{R1\ R2\ C1\ C2}} \tag{3}$$

Figure 4 explains the single-phase inverter response included PWM pulses waveform also.

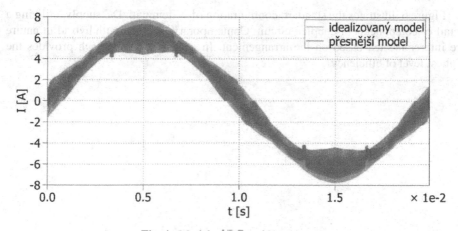

Fig. 4. Model of DC to AC converter

Figure 5 presents an overview of the key plots employed to simulate the operation of inverters, which power a single refugee shelter. Additionally, this inverter is equipped with a battery storage system that is connected to the supply grid.

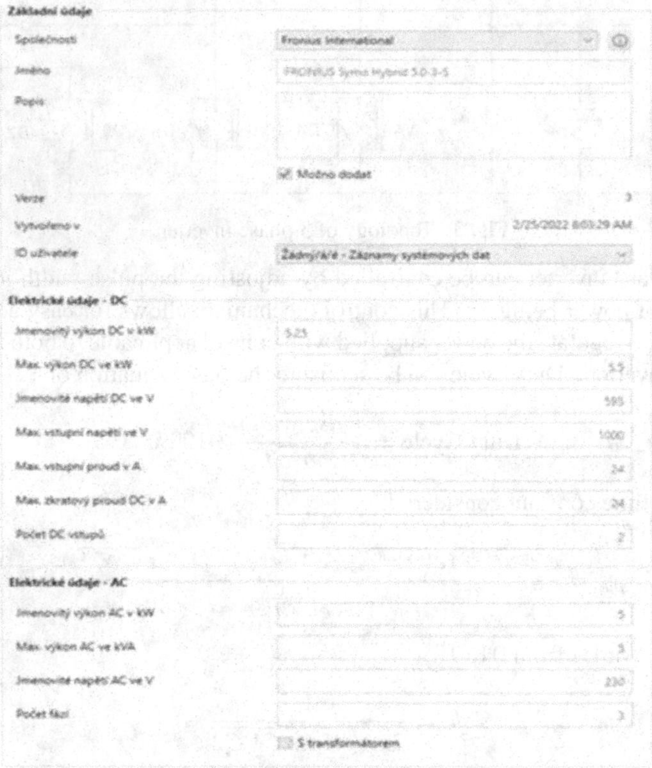

Fig. 5. Inverter Parameters

Figure 6 illustrates the standard configuration of system with DC supply, utilizing a detail of components of coupled circuit. Contemporary inverters with hybrid in nature are interconnected based on the arrangement. In general, this approach provides the highest level of efficiency.

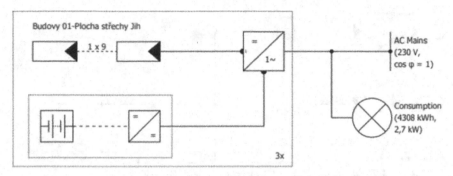

Fig. 6. Internal circuit coupled system – DC supply.

Figure 7 This text showcases the distinction of next topology in systems coupled with DC supply. In this system, the inverter of the battery is coupled in a single line. While this may be an unconventional approach, it offers numerous benefits including a flexible topology.

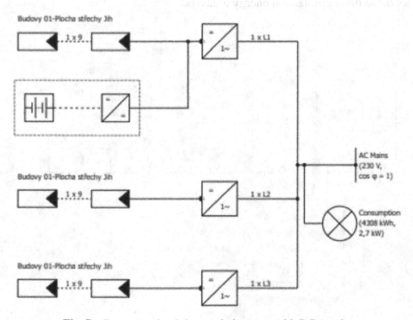

Fig. 7. Generator circuit in coupled system with DC supply.

Figure 8 illustrates the final feasible link of the inverter, which represents the simplest AC coupled system. Although this configuration is expected to have the lowest efficiency in theory, its primary benefit lies in its highly adaptable integration of different couples' system components.

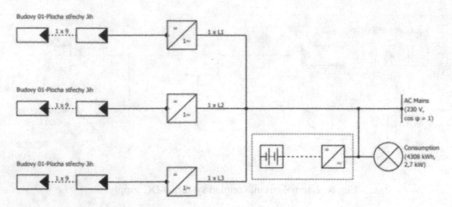

Fig. 8. System with AC coupled.

Figure 9 illustrates the necessity of minimization of energy. The chart considers various factors such as demand plot optimization, grid interconnection w.r.t energy, costs and etc. towards the grid. It ensures impact on power quality. However, this paper does not delve into optimization energy analysis.

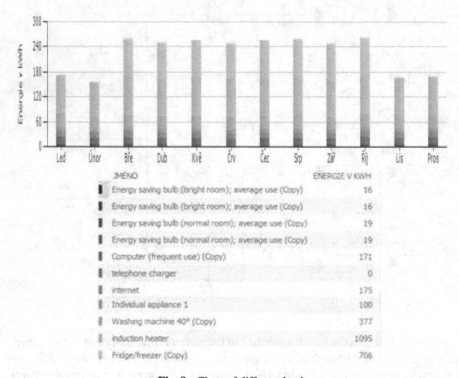

Fig. 9. Chart of different loads

3 Photo Voltaic System Model with AC and DC Coupled

Different energy gains and behaviors were observed in simulations conducted with specified objects that simulates for Vinnytsia region in emergency housing. It is important to note that these findings cannot be applicable for all to the specific conditions in Fig. 9 illustrates the disparity in different scenarios (Fig. 10).

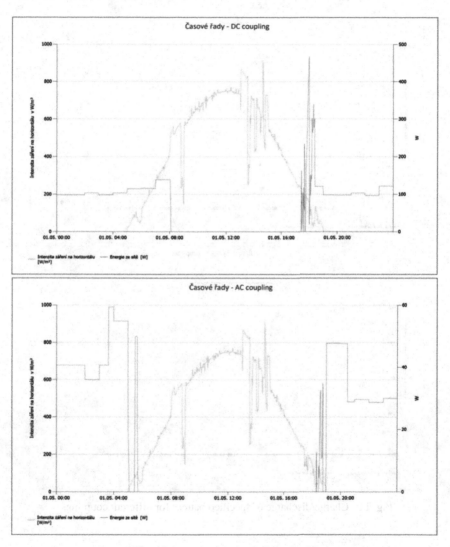

Fig. 10. Energy for simulation of grid

The intriguing aspect lies not only in the varying trajectory, but also in the temporal displacement of the withdrawals. Figure 11 illustrates the disparity in state of charge

(SOC) of the batteries under identical circumstances, while Fig. 12 showcases the losses incurred during the charging and discharging stages. It is worth noting that the significance extends beyond the mere progression of individual quantities, encompassing the commencement and conclusion of time intervals within each cycle phase, as clearly depicted.

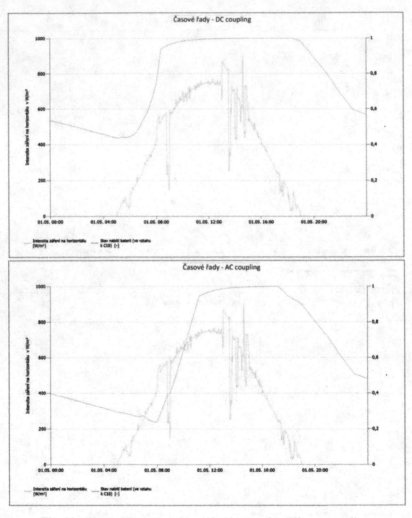

Fig. 11. Charge/discharge of specified battery for different couplings

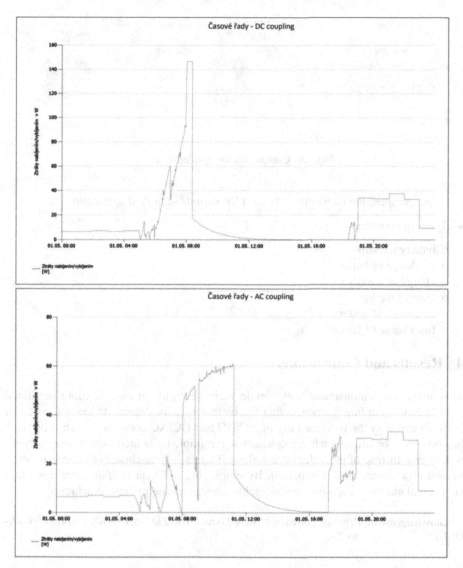

Fig. 12. Losses – AC & DC coupling Charging / discharging

Figure 13 illustrates the contrast between the output of the energy between AC and DC coupling system. It is an option that provides a higher energy yield, yet it consumes low battery power and consequently contributes more to the network. The findings clearly indicate that despite its seemingly lower conversion efficiency, the AC coupling system proves to be more advantageous in terms of overall energy utilization.

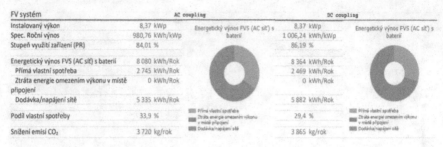

Fig. 13. Comparison of coupled systems

If we compare the latest methods used for various identified applications like:

- Types of couplings

  o battery backup
  o discharge of battery
  o charge of battery
  o state of the battery
  o discharge of battery
  o fast charge of battery

4 Results and Conclusions

This study has demonstrated with certain main scenarios, it may be more beneficial to utilize AC coupling for connecting the battery system, despite its lower theoretical performance w.r.t the increased qty of AC/DC and DC/AC conversions. The approach has proven to be original when implementing multiple units identified in war refugees of Ukraine. Instead of individual stand-alone shelters, a large cluster of these structures, known as a shelter city, is expected. By employing different coupling strategies in a substantial number of shelters, considerable savings of energy will be achieved.

Acknowledgement. Paper supported by MSCA4Ukraine ID 1233365, 23-PKVV-011, 23-PKVV-UM-11.

References

1. IEA: Renewable electricity generation by region and scenario 2018–2040 (2021)
2. Rubanenko, O.: Energy consumption optimisation of emergency shelters for Ukrainian war refugees. In: Renew. Energy Power Qual. J. (2022)
3. Belik, M., Rubanenko, O.: Degradation of monocrystalline PV panels differences between Ukrainian and Czech conditions. In: Proceedings of the 11th International Scientific Symposium ELEKTROENERGETIKA 2022, pp. 38–42 (2022)
4. Hasanpor, P., et al.: Optimum operation of battery storage system in frequency containment reserves markets. In: IEEE Trans. vol. 11, pp. 4906–4915 (2020)

5. Kazmiur, O., et al.: Determination of optimal transformation ratios of power system transformers in conditions of incomplete information regarding the values of diagnostic parameters. In: Eastern-European Journal of Enterprise Technologies (2017)
6. Belik, M.: Optimisation of energy accumulation for renewable energy sources. Renew. Energy Power Qual. J. **19**, 205–210 (2021)
7. Belik, M., Rubanenko, O.: Implementation of digital twin for increasing efficiency of renewable energy sources. Energies **16**(12) (2023). https://doi.org/10.3390/en16124787
8. Dashtdar, M., et al.: Protection of DC microgrids based on differential protection method by fuzzy systems. In: 2021 IEEE 2nd KhPI Week on Advanced Technology, KhPI Week 2021 - Conference Proceedings (2021)
9. Lezhnyuk, P., et a.: Information support for the task of estimation the quality of functioning of the electricity distribution power grids with renewable energy source. In: 2020 IEEE 7th International Conference on Energy Smart Systems, ESS 2020 (2020)
10. Belik, M.: Weather dependent mathematical model of photovoltaic panels. In: Renew. Energy Power Qual. J. (2017)
11. Komar, V., et al.: Determination of similarity criteria in optimization tasks by means of neuro-fuzzy modelling. In: Przeglad Elektrotechniczny, vol. 93 (2017)
12. Yanovych, V., et al.: Analysis of instability generation of Photovoltaic power station. In: 2020 IEEE 7th International Conference on Energy Smart Systems, ESS 2020 – Proceedings (2020)
13. Belik, M.: Simulation of photovoltaic Panels thermal features. In: Proceedings of the 18th International Scientific Conference on Electric Power Engineering, EPE 2017 (2017)
14. Hunko, I., et al.: Influence of solar power plants on 0.4 kV consumers. In: 2019 IEEE 60th Annual International Scientific Conference on Power and Electrical Engineering of Riga Technical University, RTUCON 2019 – Proceedings (2019)

A Manual Charging Adaptive Energy Efficient Bike

Harivardhagini Subhadra[1]([⊠]) [iD], V. Sreelatha Reddy[1] [iD],
and Pranavanand Satyamurthy[2] [iD]

[1] CVR College of Engineering, Mangalpalli, Telangana, India
harivardhagini@gmail.com
[2] VNR Vignana Jyothi Institute of Engineering and Technology, Hyderabad, Telangana, India
pranavanand_s@vnrvjiet.in

Abstract. An adaptive bike prototype that employs a new manual charging mechanism to generate energy from renewable sources is the primary focus of this research. This study presents a sustainable method for short distance travel in response to the growing pollution from vehicles and the rising need for environmentally friendly modes of transportation. Proving that human power can effectively replace non renewable energy sources is the main goal of this study. By incorporating a smartly engineered adaptive bicycle, our goal is to transform the mechanical energy of rotation into electrical energy, which can then be stored in a dedicated battery. After that, the bike's electric motor draws on the stored energy to propel the rider forward. Building and testing a prototype that efficiently collects and stores energy while pedalling is an important part of our technique. A thorough evaluation of the mechanical parts, the method of electrical conversion, and the effectiveness of battery storage are all part of the process. We test the adaptive bike's ability to charge itself without any outside power sources by conducting a battery of controlled trials and performance assessments. With an average efficiency of X% in energy conversion, the data show that the manual charging method was successfully integrated. Adaptive bicycles, in particular, show great promise for widespread use as a green, efficient mode of transportation for shorter commutes. This fresh perspective emphasizes the possibility of incorporating renewable energy technologies with conventional transportation, encouraging a greener and more long lasting way to get about. The accomplishment of including a manual charging mechanism for bicycles is a noteworthy new discovery that enables sustainable and environmentally friendly short distance riding. Reducing reliance on non-renewable energy sources for transportation is one of the environmental challenges that this invention aims to solve.

Keywords: Pedaling · Electric Bicycle · Manual Charging · Energy · Battery · Non Renewable resources

1 Overview

1.1 Introduction

Transportation is crucial to contemporary life in many ways, and smooth operations of transportation networks are necessary to keep economies from collapsing. There has been a marked increase in pollution due to the extensive use of petrol and fuel in traditional

© The Author(s), under exclusive license to Springer Nature Switzerland AG 2024
S. L. Gundebommu et al. (Eds.): REGS 2023, CCIS 2081, pp. 148–159, 2024.
https://doi.org/10.1007/978-3-031-58607-1_11

cars. As a solution to this pressing problem, more and more research and development is being devoted to creating eco-friendly, self-propelled electric bicycles. As petrol prices continue to rise and environmental concerns continue to grow, more and more people are turning to bicycles as a reliable and sustainable form of transportation.

Renewable energy has as of late gotten a ton of consideration because of the developing number of people in the world and the fact that fossil fuels are becoming more scarce and harmful to the environment. At the same time, pollution has become an urgent problem on a worldwide scale. When properly harnessed, human power offers a viable alternative to fossil fuels and other non-renewable energy sources. One of the most practical and environmentally friendly ways to power various low energy gadgets is using pedal power. It has a wide range of uses in transportation, both on land and at sea, and in exercise machines including ergometers, stationary cycles, steppers and elliptical trainers. Incorporating pedal power into different equipment has several benefits, including promoting physical fitness and reducing environmental impact. This leads to a healthier and more sustainable environment overall. When it comes to solving climate change, electric bicycles may be game changers. But how affordable and useful they are will determine how widely used they are. Rectifiers are the standard method by which the batteries of conventional electric bicycles are charged from the mains power source. This method makes them reliant on other power sources and reduces their utility while charging. Our goal in this study is to create an adaptive bicycle that can operate on renewable energy sources that are produced by the pedaling of its riders. This new invention has two uses: first, it's a greener way to go about, and second, it encourages exercise, much like a fitness bike.

1.2 Objective

Powered by a motor and a battery, electric bicycles provide a wonderful substitute for conventional bicycles. There are many of electric bikes on the market, but they all need an external power source to charge. With that in mind, the concept of a cheap electric bike that could be charged by pedaling was born. In order to power the bike, we are primarily concerned with converting the energy generated by pedaling into a form that can be stored in a battery. An affordable, sustainable, and environmentally friendly electric bicycle that can be manually charged using renewable energy sources is the driving force behind this project. These may be easily installed in homes, gyms, and other locations to convert and store the energy generated by riding a bicycle. This way, the energy is not lost and can be put to good use.

1.3 Motivation

Preserving natural resources and ensuring their availability for future generations is greatly aided by sustainable resource usage. Given the persistence of pollution and other environmental problems, it is critical to address them and mitigate their impact. Everyone is aware that emissions from cars and other vehicles contribute significantly to air pollution. With this in mind, we have come up with a concept that might replace these cars; it is pollution free and environmentally good as it doesn't run on petrol. Creating a versatile bicycle that can be used as an alternative to a fitness cycle, recharges itself via free pedaling, is simple to use, and handy for everyone is our key motive for this endeavor.

2 Literature Survey

Vivek et al. [1] offered a device that augments the riding experience of regular bicycles by integrating an electric engine and alternator. Especially on difficult terrain, this system's ability to switch between electric and manual modes makes riding easier. On rough terrains, you may engage the battery powered motor; on level ground, you can pedal manually. As gasoline costs continue to rise and more people grow concerned about the environment, bicycles have emerged as a popular green transportation option. Bicycles are already a great mode of transportation, but this initiative intends to make them even more accessible by lowering the amount of physical effort needed to ride one. The main objective is to support environmentally friendly modes of transportation, with an emphasis on university settings, in honor of "GREEN ENERGY." You may get specifics on the weight of the item in the report.

Shubham et al. [2] unveiled a new electric bike that can power itself without any help from outside sources. The dynamo and PMDC motors are integrated into the crank mechanism of this bike, which allows it to propel the back wheel. When turned on, the motor uses 12 V and 14 amps from two drycell batteries that are linked in parallel. Simplified, effort free riding on flat and sloping ground, and enhanced performance on tough terrain are both made possible by the action of the engine, which engages a flywheel, which powers the bike's back wheel via a sprocket system. Even though they are essential for getting about in today's fast paced society, automobiles and motorcycles add to pollution levels due to the burning of gasoline. We have limited fuel supplies, and this pollution is making things worse for the environment. As a result, solutions for sustainable travel are becoming more important. One form of this technology is the electric bike, or E bike, however it isn't without its flaws. The development of the self-generating electric bike was a response to these constraints. Thanks to its revolutionary design, this device can function independently of gasoline or external battery charge. Bike functioning is unaffected and environmental effect is reduced since it produces its own electricity.

Sunil et al. [3] put forth a technique for producing electricity by means of pedalling a bicycle, which would transform mechanical energy into electricity. They tested the viability and efficiency of this bike generator in a gym. Various parts of the system were assembled, including a battery, generator, bicycle, sprocket, inverter board, generator stand, belt, and chain. By pedalling it by hand, the bicycle generator was able to produce 19.5 Wh/h of electricity. The application of force to the pedal started the rotation of the main sprocket, which was linked to a smaller sprocket at the back by a chain. The rear wheel turned because this rotation reached the rear centre shaft. A DC generator might transform mechanical motion into electrical energy by connecting the rear wheel to it via a flat belt. The direct current (D.C.) that was produced could be stored directly in a battery. To make the direct current (D.C.) usable in the supply port, an inverter board was used to transform it into alternating current (A.C.). The viability of power production by bicycle was thoroughly investigated, the working principle of the generator was explored in detail, and many cost estimations pertaining to profit and loss were manually computed as part of the task.

Manish Yadav et al. [4] Go further into the paper's central idea, which is to connect a bike to a stationary device that can turn the rider's pedalling effort into usable electricity. There are two parts to the system that make up the mechanism that turns mechanical energy into electrical energy. The electric block transforms the mechanical block's energy into electric energy, while the mechanical block adjusts the pedals' rotating movement to the generator's needs. A dynamo or an alternator may transform the mechanical energy of pedalling into electrical energy. Cause. This research introduced an energy scavenging system that uses recycled and autonomous components to capture energy that is used during physical activity.

Reddi Sankar et al. [5] talk about a bike that uses pedal power in addition to solar electricity. An electric motor (DC motor) powered by solar panels attached to the bike's frame powers this adaptation of conventional bicycles. At the point when the bicycle isn't being used, these panels charge the battery and power the hub motor. The front axle of the solar assisted bike has a 250W DC hub motor, allowing it to reach speeds of 25–30 kmph. A 20 W photovoltaic solar panel, two 35 Ah lead acid batteries, an accelerator, a 24V 10 Amp voltage regulator, and a 24V 25 Amp motor controller are all part of it. In the event that cloud cover prevents enough solar power, it also has a 220-240V AC wall outlet for charging the batteries. With a cost of around Rs. 0.70/km, operational expenses are minimal. This model is simple to set up and needs little upkeep. Whether you're riding on cement, asphalt, mud, or any other sort of road, this solar assisted bicycle modification can handle it. People who travel small distances often use it, such as schoolchildren, college students, workplace commuters, peasants, and postmen, since it is inexpensive and easy to build. It helps the impoverished and welcomes people of all ages and capacities. Free of charge, it runs all year.

Megalingam et al. [6]. Offered the idea to get a better understanding of how pedalling generates power. The dynamo, which transforms mechanical energy from pedals into electrical current, is used in this work. But only low power gadgets may be charged using dynamic. Modern technology has rendered dynamos obsolete due to the low output power they provide. Many Indian villages rely on bicycles as their primary means of mobility. A large portion of these towns lack access to electricity. A dynamo or alternator may transform the mechanical energy of pedalling into electrical energy. Nighttime study sessions may be enhanced with the help of small, powered illumination devices that can be charged using dynamos. You may use this same idea to charge your cell phone, iPod, laptop, etc. It is also possible to produce extra power by turning the wheels of alternator vehicles, such as motorcycles and automobiles. One option is to utilise the electricity in the same vehicle, while another is to store it in a battery and use it to power other equipment. As an added bonus, you can get some exercise and build some power just by riding a bike. In this article, we will look at several ways that you can ride your bike to generate power. Additionally, it provides a thorough explanation of how to create electricity using a bottle dynamo. Additionally, a comprehensive evaluation of pedal power is offered.

Panicker.et al. [7]. Explored the concept of a hybrid electric bike, which charges its battery using a combination of solar panels, pedalling, and an AC power source. Making a method that can recharge itself is the goal, as is reducing or doing away with the need to plug into an external power source entirely. Presented a three way charging electric bicycle in this article. While riding, a dc generator generates electricity. A photovoltaic cell produces energy even when the sun isn't shining. We charge the battery using a charger that plugs into the wall outlet. The electric bike's 24V250W brushless direct current (BLDC) hub motor and lithium ion (Li ion) battery provide the necessary power. A BLDC hub motor eliminates the inefficiencies and complexity of a permanent magnet direct current (PMDC) motor, making it ideal for bicycle applications. These days, electric bicycles don't use sealed maintenance free (SMF) batteries; instead, they use Li ion batteries.

Belekar et.al. [8] The suggested task was to develop and evaluate a self-charging system for a battery powered motorcycle in order to achieve greater energy efficiency. A battery electric vehicle (BEV) regeneration system is an attempt to maximize the power of the electric motor and recharge the batteries using energy from the wheels' movement. To accommodate the self-charging technology and different sized batteries, the chassis of a motorbike that is already on the market is modified. Because of how everything was put up, the chain sprockets could send the rotational energy they experienced to the alternator via the motor and DC-DC converter. The 14.4V DC result of the alternator is provided to the DC converter here by means of a battery supply. This DC-DC converter boosts the input voltage to 54V, which is all that could possibly be needed to charge four batteries in series and give 48V of usable power. Consequently, the motor's shaft is powered by the batteries, which get sufficient electricity to recharge. Using a multimeter, we measure the distance driven with and without the recharging circuit to determine how well the vehicle's source supply charges the batteries.

Prashant Kadi et al. [9] One new feature of hybrid powered electric bicycles is their revolutionary charging mechanism, which integrates solar power, a dynamo, and a regular 220V AC wall charger to recharge the batteries more effectively than ever before. All of these things work together to power an electric PMDC motor, which in turn propels the bike. Users have the freedom to effortlessly transition between two separate modes of operation thanks to its adaptable design. This ecofriendly and adaptable mode of transportation gives riders the option to pedal or use the electric PMDC motor for power.

Yogesh et al. [10] designed a brand new idea: the E Bike that can charge itself. Because of the lack of a robust charging infrastructure, traditional electric bicycles suffer from short battery lives and lengthy charging periods. This study offers a revolutionary solution by suggesting a plan that permits the E Bicycle to charge its battery while riding. The E Bicycle's reach is much increased and the need to charge it outside is decreased since it switches to a rechargeable battery when the first one is empty. E Bikes, made possible by this novel method, improve in utility and efficiency while being able to go greater distances than traditional, fuel powered motorbikes. A potential answer to the problems with existing E Bikes, the Self Recharging EBike has a longer range on a single charge while maintaining the same battery capacity.

Nagaraj et al. [11] have put out an innovative approach to overcome the conventional constraints of electric bikes, with an emphasis on the pressing issues of battery life and charging durations, which are especially pertinent to the situation in which India's infrastructure is developing. Their revolutionary self-recharging electric bike concept is a giant stride forward, thanks to its electric motors and electrochemical batteries. They made E Bikes more competitive with gas powered bikes by integrating a smooth battery switching system, which greatly increases the bike's range while decreasing the need for external charging stations. An ecofriendly new form of transportation that might shake up the electric vehicle market, the Self Recharging E Bike basically doubles the range of conventional bikes while maintaining the same battery size.

Priyanka et al. [12] investigated the possibility of creating a green electric bike that uses a renewable energy source. With a focus on four key areas—motor efficiency, decreased battery charge time, cost effectiveness analysis, and design optimization— their method expertly transforms mechanical energy into electrical power. An electric bike's suggested regeneration system includes a lightweight frame, a controller, an alternator, a lithium ion battery pack, a brushless DC motor, and other essential parts. As a whole, this design has the makings of a hybrid battery system that might increase performance and prolong the life of the battery. Their research shows that the alternator's ability to generate power is highly dependent on the BLDC hub motor's speed and torque. Particularly when the motor is running at 2900 rpm, 48 V, and 32 amps, this system offers a significant and encouraging alternative to traditional E bike commuting while significantly reducing its negative effects on the environment.

3 Implementation

This adaptive bike may be powered entirely by the energy you use while pedaling freely; there's no need to plug it into an outlet. The hub motor then runs the cycle by drawing on the electricity that has been charged in the battery. The result is efficient utilization of energy. This particular kind of alternator relies on a BLDC hub motor, battery, and motor controller as its primary components. Power is transformed from mechanical to electrical form and then stored in a battery by means of the alternator. Energy may be transformed from electrical to mechanical with the help of the BLDC Hub motor. You may adjust the speed of the Center point engine utilizing the engine regulator. The creation of a versatile bicycle that is easy for everybody to ride is our top priority. An environmentally beneficial and clean kind of power is pedal power. Rather of relying on fossil fuels or an electrical grid, this prototype generates its own power by pedaling, a sustainable energy source. So, it's good for the environment and helps save power. People who aren't able to ride a regular exercise bike may find an adaptable bike to be an encouraging and practical substitute.

3.1 Block Diagram

As seen in Fig. 1, the charge mode primarily consists of the three components pedaling, alternator, and battery, which store energy. This schematic depicts the charging method when the battery is charged by free pedaling by hand. On a motorcycle, the alternator is connected to the back wheel. One tool for transforming mechanical energy into electrical energy is the alternator. The wheel begins to spin as we press down on the pedals, which causes the alternator that comes into touch with the wheel to likewise begin to spin.

Electrical energy is generated from rotational energy by means of the alternator. A power source known as a battery receives the electrical signal produced by the alternator and stores it for later use. So, as you cycle freely, an alternator charges the battery.

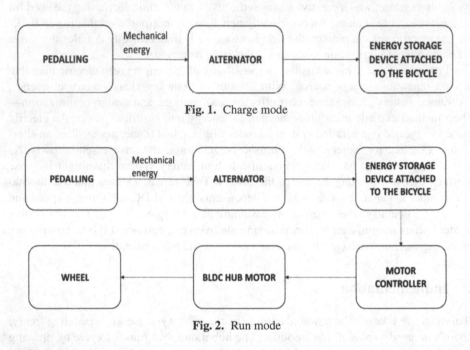

Fig. 1. Charge mode

Fig. 2. Run mode

There is a run mode shown in Fig. 2. The motor controller is linked to the battery and the BLDC hub motor. The BLDC Center engine, which changes electrical energy into mechanical energy, is connected to the bicycle's rear wheel. The hub motor here makes use of the energy that the battery stores while you pedal (Charge mode). The energy that comes from the battery is transformed into mechanical energy, namely rotational energy, by the hub motor, and then it begins to spin. The outcome is that the wheel spins, propelling the bike forward.

3.2 Schematic Diagram

The connection between the alternator and the battery is shown in the schematic design in Fig. 3. At the back of the bike, you'll find the alternator fastened with a belt. You may connect the alternator's output terminals to a DC step up booster. Two wires go from the battery to the step up booster's output terminals.

Figure 4 displays the connections for the run mode. The motor controller has primary control over the hub motor. There is no misalignment between the battery connectors and the motor controller. With the correct wiring, all the other inputs and outputs, such as the throttle, brakes, BLDC Hub motor, etc., are linked to the terminals of the motor controller unit. At last, the bicycle's unique wheel receives the BLDC Hub motor.

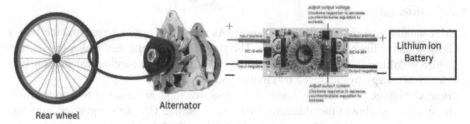

Fig. 3. Charge mode Schematic Diagram

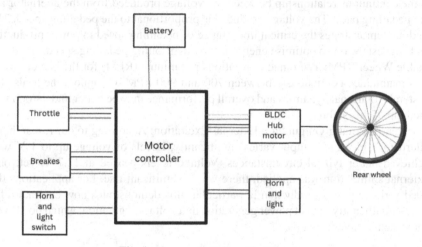

Fig. 4. Run mode Schematic Diagram

4   Results and Conclusion

Turning mechanical energy into electrical energy, storing it in a battery, and then using this stored power to move the bike forward is the fundamental function of an adaptive bicycle. Because of this, the process consists of two steps: charging and operating.

4.1   Step One (Charge Mode)

An alternator is being used to charge the battery in this set up. Ensure that the alternator is securely fastened to the bicycle's back wheel and maintained in a snug fit. The back wheel's revolution causes the alternator, which is in direct contact with the wheel, to likewise revolve. This mechanism transforms mechanical energy into electrical energy [13]. In the end, the battery is charged via this electrical signal. Electricity is generated from mechanical energy, often called rotational energy, and this is how the energy is kept in the battery [14].

4.1.1 Observations

a. Mechanical Integration with Rear Wheel: As set up for our experiments, the bike's back wheel is linked to the alternator. The foundation for efficient power production while pedalling is this direct mechanical link. It converts mechanical motion into electrical energy by capturing its kinetic energy.
b. Precise Voltage Monitoring: Accurate voltage readings are assured by careful multimeter monitoring of the alternator's electrical output. Our data is guaranteed to be accurate and reliable thanks to this methodical methodology.
c. Pedaling Speed and Voltage Correlation: One remarkable finding from our research is the continuous relationship between the voltage produced from the alternator and the pedalling pace. The voltage produced is proportional to the pedalling speed. This finding demonstrates the critical importance of pedalling speed in power production and suggests ways to optimise energy use by controlling pedalling speed.
d. Stable Wheel RPM: The range of rotations per minute (RPM) for the bicycle wheel was maintained continuously between 700 and 900 RPM throughout the trials. The system's rotational dynamics and overall performance may be better understood with the help of this stability.
e. Variable Alternator Output with External Excitation: According to our research, the alternator's electrical output varies significantly. Levels of voltage up to 1 V were achieved under typical circumstances. Voltage levels peaked at 12.7 V, yet, once external stimulation was applied, there was a significant rise. For applications that need variable power production in particular, this demonstrates how external influences significantly impact power generation and voltage regulation, which bodes well for system efficiency gains.

4.1.2 Limitations

The prototype's alternator, a sort of synchronous generator, is the subject of a crucial note. To produce a fair and consistent output, it is important to note that external stimulation is required. This suggests that the alternator may not work as well or as dependably if not stimulated outside.

A major drawback becomes apparent when we think about how insufficient the boosted output voltage is for adequately charging the battery, even when voltage is produced at the alternator output. Because of this shortcoming, we worry that the prototype won't be able to keep the battery supplied with enough power, which might reduce its total performance.

4.2 Phase – 2 (Run Mode)

We transform the bike into an electric bike at this stage (called "run mode") by connecting it to a battery and a BLDC hub motor, which acts as a motor controller. To begin, connect the BLDC Hub motor to the bike's back wheel. The energy generated by pedalling or cycling is stored in a battery that is linked to this motor. To regulate the motor's speed, a motor controller is used. To propel the bike, one uses the energy stored in the battery to turn the BLDC hub motor.

4.2.1 Observations

If you turn the key and then crank up the throttle, the bike will start to move forward because the BLDC Hub motor starts to spin. Indicative of a responsive and operational system, this finding highlights the motor's effective participation in pushing the bicycle.

The throttle is the primary control lever for the bicycle's speed as it has a direct effect on the motor's output. You may have complete command of the bike's pace thanks to the brakes, which also serve to stabilise it. This observation demonstrates how the system is both flexible and easy to handle, allowing the cyclist to speed up or slow down depending on the situation. The bike is carrying a total of around 4.5 kilos.

4.2.2 Limitation

The bike didn't seem to have any problems getting about, although it was carrying around about 4.5 kg. If you want to know how well the bike performs in terms of things like speed, energy efficiency, and rider comfort, you have to take this extra weight into account. All things considered, the bike's performance and capabilities should be evaluated with this weight in mind.

Fig. 5. Prototype of Adaptive bicycle

Fig. 6. Alternator attached to the Rear wheel.

The cycle's prototype is shown in Fig. 5. The alternator, as shown in Fig. 6, is attached to the bicycle's rear wheel. One of the most common ways to get about and go to work is in a vehicle. Their contributions to the economy are substantial, and they have also made our lives easier. The fact that these vehicles' emissions do significant damage to the environment is, nonetheless, an undeniable truth. In addition, there is a risk that non renewable resources may become scarce if current consumption patterns persist. This has brought attention to the need for transport technology that are less harmful to the environment. Because they are both inexpensive and environmentally friendly, bicycles are the best option for getting about town. An efficient and environmentally friendly electric bicycle is the primary focus of our effort. Using renewable energy sources and hand charging as its foundation, this project creates a prototype of an adaptive bicycle.

By converting the rotational force into electrical form and storing it in the battery, the power generated by pedaling may be used to propel the bicycle forward. Hence, the bike may be charged wirelessly without an external source since the battery is charged manually while riding. This kind of pedal assist electric bike is a great replacement for fossil fuels since it makes effective use of human energy to charge the battery. Since this bike uses less electricity and produces less pollution, it can also be economical and good for the environment. Alternatively, this electric bike has the potential to encourage people to maintain a healthy lifestyle and be more active by making better use of the power needed to charge the bike. Since it may be converted into an electric bike and used as a substitute for a training cycle, this manually charged bike might be useful for ensuring the sustainable use of energy.

References

1. Kumar, V.V., Karthik, A., Roshan, A., Kumar, A.J.: International Journal of Innovative research in Science, Engineering and Technology. In: Design and Implementation of Electric Assisted Bicycle with Self Recharging Mechanism, vol. 3, Special Issue 5 (2014)
2. Tayde, S.U., Makode, N.W., Laybar, U.M., Rakhonde, B.S.: Self Power Generating Electrical Bicycle, vol. 04(01) (2017)
3. Sunil, J., Sathiya Priyan, B., Anish Anto, L., Muthuvel, L., Arun Daniel, T.: Bicycle Power Generation and Its Feasibility. Research Gate publications (2015)
4. Yadav, S.M., et al.: Power generation using bicycle mechanism as an alternative energy source. Int. Res. J. Eng. Technol. (IRJET) 05(04) (2018). e- ISSN: 2395–0056
5. Sankar, M.R., Pushpaveni, T., Reddy, V.B.P.: Design and development of solar assisted bicycle. In: Int. J. Sci. Res. Publ. 3(3) (2013)
6. Megalingam, R.K., Veliyara, P.S., Prabhu, R.M., Katoch, R.: Pedal power generation. Int. J. Appl. Eng. Res. 7(11), 699–704 (2012)
7. Panicker, B.M., Sajeev, A.P., Akhil, P., Babu, A.R., Arjun, K.U., Varghese, N.: Electric bicycle with three way charging. Int. J. Adv. Res. Electr. Electron. Instrument. Eng. 5(4) (2016). (An ISO 3297: 2007 Certified Organization)
8. Belekar, R.D., Subramanian, S., Panvalkar, P.V., Desai, M., Patole, R.: Alternator charging system for electric motorcycles. Int. Res. J. Eng. Technol. (IRJET), 04(04) (2017). ISSN:2395–0056
9. Kadi, P., Kulkarni, S.: Hybrid powered electric bicycle. IJSRD – Int. J. Sci. Res. Dev. 4(05), (2016)
10. Jadhav, Y., Kale, G., Manghare, S., Patil, S.: Self-charging electric bicycle. Int. Journalon Recent Innov. Trends Comput. Commun. 4 (2017). ISSN: 2321–8169
11. Sindagi, N., Nandan, K., Sunil, N., Sumanth, H.R., Dhanalakshmi, R.: Design and analysis of self recharging electric vehicle. Int. J. Res. Eng. IT Soc. Sci. 10(06), 44–51 (2020). ISSN 2250–0588, Impact Factor: 6.565

12. Arora, P.G., Sagor, J.A.: Design approach of regenerative system for an electric bike using alternator. JETIR **5**(12) (2018)
13. Vinay, J.K., Kumar, P.A., Rajput, C., Saxena, S.: Electric bicycle battery charging estimation while pedaling. Int. J. Creat. Res. Thoughts (IJCRT) **10**(5) (2022)
14. Sajid, M.D., Quadri, M.U., Khan, A.A., Ayjaz, S.: Power generation through pedaling. Grenze Int. J. Eng. Technol. Special Issue, Grenze Scientific Society (2018)

Review of Optimization Tools Used for Design of Distributed Renewable Energy Resources

Muthukumaran Thulasingam$^{(\boxtimes)}$ ⓘ and Ajay-D.-Vimal Raj Periyanayagam ⓘ

Department of Electrical & Electronics Engineering, Puducherry Technological University, Puducherry, India
amtechhy@gmail.com

Abstract. This paper going to review the different optimization tools used in the distributed energy systems. Optimization tools are generally used to size the individual renewable sources.as well as used to analysis the economic part of the system in terms of cost of Energy (COE), Net present cost (NPC), Internal rate of return (IRR), simple payback period, Operating cost and capital cost of system. Due to rapid development in technology, various optimization tools such as Genetic Algorithm, Machine Learning Algorithm, Fuzzy logic, Artificial Intelligence was developed by the researcher for optimize hybrid renewable energy system (HRE). Because of global warming, penetration of renewable energy in the grid, commercial building, educational building and residential consumer was increasing gradually and it is very important to have a optimization tool to check the feasibility of the project before going for the real time installation of HRE system. In this paper various optimization tools used by the researcher is going to be investigated in detailed.

Keywords: Optimization tools · HRE · Economic Analysis

1 Introduction

Because of civilization in the country, energy requirement for individual need is increasing in nature. Energy produced from the existing the conventional energy is not quite enough to meet the energy requirement / demand across the customer in different domains which includes domestic consumer, industrial consumer, commercial building, agricultural sectors and educational institute. Because of penetration of renewable energy in our grid [1]. On year on year contribution from the renewable energy is increasing gradually in our grid energy. Especially growth of solar PV generation is increasing tremendously, both private and government sector contribution is more, India had total solar PV system installation of 48.65GW as of Nov 2021. Integrating the renewable energy in to grid is challenging task and output from renewable sources especially solar and wind is intermittent in nature. Before the optimizing the size of renewable energy resources, various factors need to be considered such as load type, peak load, peak load time duration, off peak load, off peak load duration, etc. [2]. Any project to be executed, it needs to be checked for financial feasibility. Based on the outcome of financial analysis, project

© The Author(s), under exclusive license to Springer Nature Switzerland AG 2024
S. L. Gundebommu et al. (Eds.): REGS 2023, CCIS 2081, pp. 160–174, 2024.
https://doi.org/10.1007/978-3-031-58607-1_12

owner takes the call to whether to proceed for the project execution or not. Here the prediction of system performance in terms of cost becoming necessary for high valued project [3]. The most of researcher devised the different optimization technique in order to analysis the financial aspect of the project. In the market there is different optimizations software or tool are available especially to optimize the renewable energy and to study economic analysis of distributed energy system [4]. The renewable energy output is intermittent in nature due to this storage of energy is becoming essential part of renewable system. Due to the advancement in technology especially in the energy storage, cost of storage system is declining gradually [5]. The output from single renewable energy sources will not be sufficient to manage the load demand in the case of standalone system. Adding the more than one renewable source called Hybrid renewable energy (HRE) in standalone system will reduce the load demand and it will satisfy the load demand at all time [6].

2 DER (Distributed Energy Resources)

Microgrid is in place in the most of remote areas where there is problem in transferring the power through T & D network. It becomes easier to transfer the power if distribution network is nearer to the load centers. Microgrid is generally fed with the hybrid renewable energy sources like PV, Wind, Biomass, and Biogas called as DER and with the energy storage will make the system more reliable [7]. This Microgrid system fed with DER will be cost effective when compared with the system fed with grid energy supplied through T&D network. The running cost of grid energy is high, whereas running cost of Microgrid with DER will be less since the grid energy is dominated by fossil fuel. The cost of coal per ton and energy production cost per Mwh was given in Table1 [8].The renewable energy cost in India per Mwh was shown in the Fig. 1 [8].

Table 1. Cost of coal

	Cost/t		Cost/MWhth	
	2015	2030	2015	2030
Coal In India	2015	2030	2015	2030
Coal In India	30.08	63.91	3.69	7.85

Cost of renewable energy and non-renewable energy comparison with respect to COE is shown in Fig. 1.

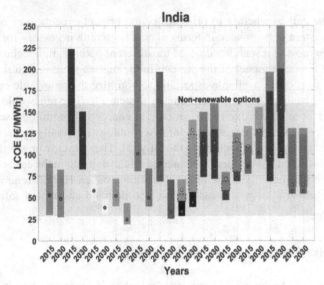

Fig. 1. Renewable Energy cost in India

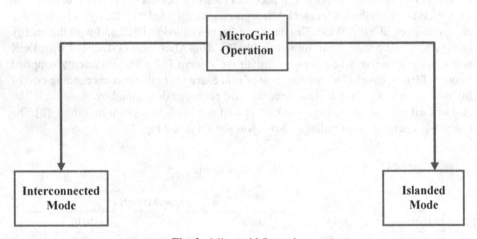

Fig. 2. Microgrid Operation

Figure 2 depicts the Microgrid operation and it can be operated in interconnected mode consist of utility grid and hydrid renewable energy sources, whereas microgrid in islanded mode consist only hybrid renewable energy sources [9].The microgrid offers advantages to the customer as well as to utility grid also,

Less carbon emission because of higher renewable penetration.

Less operating cost because of T&D cost is zero.

High Reliability.

Less power quality issue in the loads.

Better stable operation of the system.

3 Designing Tools of HRE (Hybrid Renewable Energy)

Designing tools available in the market pertaining to designing of Hybrid renewable energy sources, each tool is having own uniqueness [10]. The designing tool plays a vital role in sizing the renewable energy based on following constraints,

- load demand,
- availability of area,
- meteorological conditions,
- availability of adequate resources for generation of required power for the load demand
- Budget / fund availability.

List of tools available for designing the hybrid renewable energy is given below, (Table 2)

Table 2. Designing tools for renewable energy sources

S.NO	Name of the Tool	Objective of the Tool
1	Solar Pro	• Used for designing the solar PV system • Used for designing the Grid Interactive system
2	PV Syst	• Used for designing the solar PV system • Used for designing the Grid Interactive system, Standalone system and pumping system
3	Helioscope	• Used for designing the solar PV system • Used for designing the Grid Interactive system
4	Homer Pro	• Used for designing Hybrid renewable energy sources which includes sizing of PV, Wind, Biomass, Battery, Electrolyser & Hydro • Used for designing the Grid Interactive system
5	PV sol	• Used for designing PV system • Used to design the Grid interactive system • Used to design the standalone system

4 Optimization Tools

The different tools mentioned available in the market for designing of renewable energy is discussed in this session.

4.1 Solar Pro

The Solar Pro software is used to design the grid interactive system and it is used to size the inverter rating, size of the array and size of the string. This software takes the consideration of shadow impact on the solar panel, thereby it will forecast the performance of the system accurately. The major features of the solar pro tool is given below,

- 3D CAD simulation
- Shadow mapping analysis
- I-V characteristic study of individual array
- Import of AutoCAD file in the software
- Auto array installation
- Auto sizing of PV plant based on availability area.
- Consideration of Albedo factor
- Consideration of subsystem loss of PV plant
- Financial Analysis
- Metrological database available in the software

The author compared the actual output of 100KW PV plant with designed data of the PV plant using Solar Pro. The results obtained from the Solar Pro tools matches with actual output of the system [11]. The Solar Pro software calculates the payback period of the PV plant based on the Investment made for the project and the cost saving achieved in the project (Fig. 3).

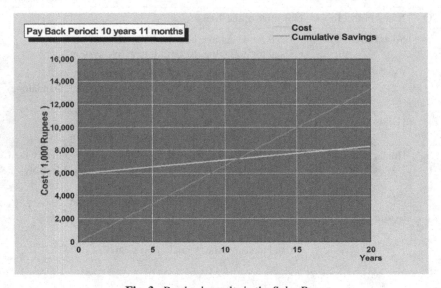

Fig. 3. Pay back results in the Solar Pro

4.2 PV Syst

The PV Syst tool used for designing the PV plant which includes the design of On grid system off grid system and pumping system. This tool is used to,

- Size the Array
- Size the string
- Size the Inverter
- Size the battery rating
- Size the Number of battery required for the project

The PV Syst is used to calculate the subsystem loss there by it will forecast the output of the PV plant accurately. This software will give clarity to designer as well as to project owner to address the losses in the system since this tool will forecast the performance ratio (PR) of the PV plant. The performance of PV plant with fixed tilt angle of PV module with different azimuth angle of the system is analyzed for educational institute load using PV Syst [12].

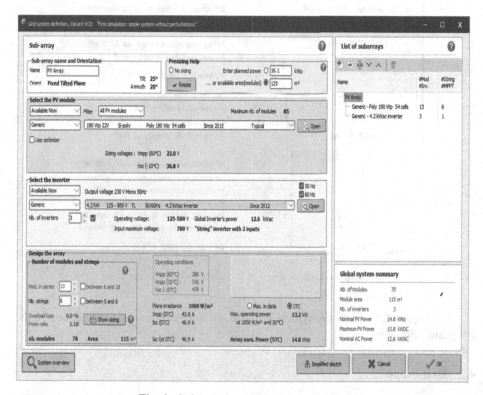

Fig. 4. String design and inverter selection

The author able to find out the performance ratio of the PV plant for the load at different azimuth angle. The 700KWp on grid system was simulated using PV Syst for the Daikundi province in the Afghanistan. The PV Syst forecasted the output and performance ratio of the 700Kwp system as 1266MWh/yr and 0.797 respectively [13]. The standalone system can also be designed and performance of the system can also be studied using the PV syst. The typical office load of about 1086.4KWH was designed and simulated using PV Syst [14]. The result show the average recorded performance ratio of the plant is 0.728, the performance ratio for the all the month in the year is forecasted by the software along with loss associated with each subsystem of the plant. The sizing of PV plant for the pumping application can be done using PV Syst, the basic procedure involved for the selection of array for typical water pumping application of about 249m3 /day was designed using the PV Syst [15].

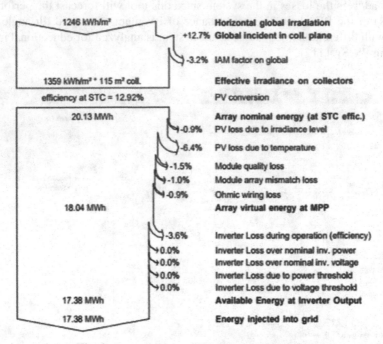

Fig. 5. Loss diagram of PV plant for whole year

The Figs. 4 & 5 shows the typical PV Syst string design and loss calculation in the PV plant [16].

4.3 Helioscope

Helioscope used for designing on grid PV system and it is web based tool. It supports 3D image file and capable of forecast the accurately the energy output from the PV plant. The major features of the Helioscope tool is given below,

- Supports unlimited design
- Capable of exporting file to CAD
- Having wide range of components in the library
- Weather data available for wide location in the world
- Supports 3D design

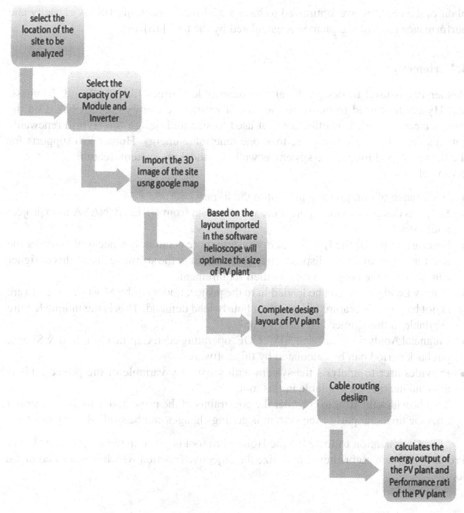

Fig. 6. Steps in designing the PV plant in Helioscope

The performance of the grid interactive PV plant was studied at different tilt using Helioscope [17] for the location Rajshahi in Bangladesh. The sample area of 20 acres of location is selected in the software and optimization of PV plant at different tilt angle was done and forecasted the energy output. The procedure for designing the PV plant in the Helioscope was shown in the Fig. 6. By using the procedure given in the Fig. 6 one can design the PV plant for any location in the world. Advantage of using the Helioscope is it will optimize the cable routing also apart from optimizing the size of PV plant and inverter sizing. This tool will help the process owner to reduce the installation cost of plant and it will do the shadow analysis of the PV plant based on output shadow mapping this tool forecast the output of the PV plant. The educational institute layout was imported in to the software and design was validated by the Helioscope. The PV plant of the campus was optimized to have a 56 Inverter each of 20KVA capacity and performance ratio of the plant was calculated by the tool [18].

4.4 Homer Pro

Homer Pro is used to design the all the renewable sources like PV, Wind, Biomass, and Hydro. It is used to optimize the size of renewable energy sources based on the requirement of load. It is effective tool used to size and design the Hybrid renewable sources i.e. combination of more than one renewable energy. Homer Pro supports for the design of grid interactive system as well as standalone system, feature of the tool is given below,

- Wide range of components present in the library.
- Supports designer to incorporate the weather data from NREL & NASA in to project under study.
- Forecast output of the Hybrid energy system more accurately since tool consider the real time input different dispatch strategy is inbuilt in the software due to this designer can optimize the project based on their requirement.
- Hourly Load profile can be loaded in to the project under study. Most of the software is not having this feature of uploading hourly load demand. This is the unique feature available in the Homer Pro software only.
- Financial Analysis in terms of NPC, COE, operating cost, Capital Cost, IRR & Simple payback period can be calculated by this software.
- Provides user to analysis the system with sensitivity variable of the project, this is also unique feature available in this tool.
- Tool provides the user to consider the constraints of the project, due to this constraint how the financial part of the system is getting changes can be studied using this tool.

The Optimization followed in the Homer Pro tool is given in the Fig. 7. The Homer Pro uses dispatch algorithm to optimize the objective function which is modelled in the tool.

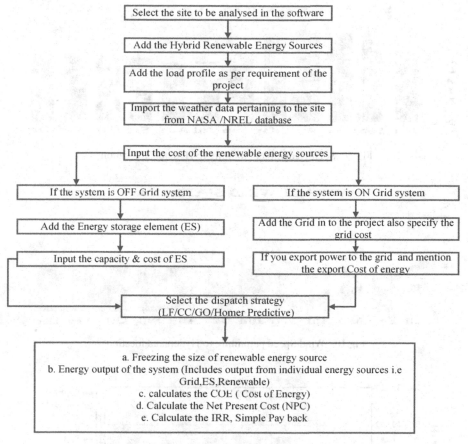

Fig. 7. Flow chart of the design steps in Homer Pro

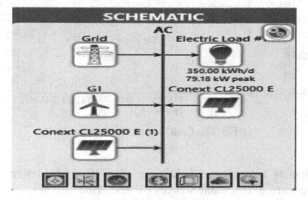

Fig. 8. Schematic diagram of energy resources

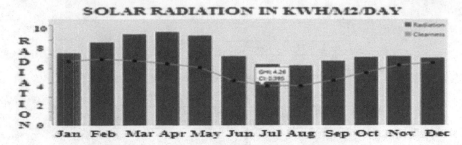

Fig. 9. Solar Radiation pertaining to Hyderabad location

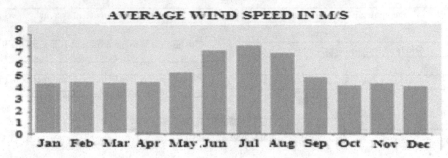

Fig.10. Wind Speed pertaining to Hyderabad location

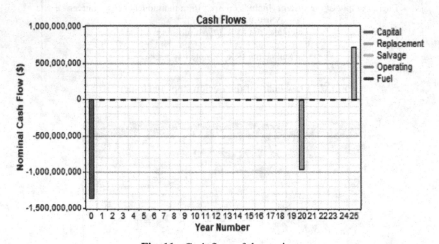

Fig. 11. Cash flow of the project.

Figure 8 depicts the schematic diagram of HRE (Hybrid Renewable Energy Configuration) consist of renewable sources PV and Wind turbines.

Figure 8 shows the schematic diagram of energy resources used in the project under study for optimization and economic analysis. Figure 9–10 depicts weather data pertaining to solar radiation and wind speed imported from NREL pertaining to particular location of the project, Fig. 11 shows the cash flow of the project [19]. Typical residential load is simulated and analyzed using Homer Pro. The peak demand of the system is 2.09 kW and average energy consumption per day is 11.26 KWH. The schematic diagram for the system is given in the Fig. 12.

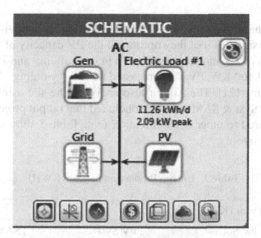

Fig. 12. Schematic diagram of residential load

The residential load requirement is currently satisfied by the grid system, renewable energy sources is added to existing system to reduce the grid energy and to reduce the carbon emission. The simulation results of the system is given in Fig. 13.

Export...						Optimization Results				
						Left Double Click on a particular system to see its detailed Simulation Results.				
Architecture					Cost				System	
PV (kW)	Gen (kW)	Grid (kW)	Dispatch	COE (₹)	NPC (₹)	Operating cost (₹/yr)	Initial capital (₹)	Ren Frac (%)	Total Fuel (L/yr)
1.00		999,999	LF	₹7.95	₹447,070	₹32,649	₹25,000	35.9	0
1.00	2.30	999,999	LF	₹8.05	₹452,551	₹32,628	₹30,750	35.9	0

Fig. 13. Optimization Results

4.5 Pv Sol

This tool is used to design the PV system it is used to design the grid interactive system and standalone system. This software having inbuilt database of PV modules of more than 17000 and Inverter database of more than 4000 nos' [20]. The features of the software is given below,

- 3D Visualization of site under study.
- 3D Preview of Sun Position.
- Auto CAD files can be imported to the software
- Cable layout Design
- Tracking of PV Modules
- Financial Analysis

The designing the grid interactive system for academic institution was carried out using PV Sol tool, using this tool they optimized the PV capacity of the institution. PV sol estimated the no of PV module of 174 no's in the building and PV plant was sized to the capacity of 59.160 KW. PV Sol forecasted the energy output of 81,838 KWH per year from the building [21]. The Author also analyzed the site using three simulation tool i.e. PV Sol, PV Syst & SAM. Author concluded that output projection from the PV Syst is more compared to other two simulation tool. Table 3 shows the energy output from three simulation tool.

Table 3. Energy Production in year (KWH)

Energy production in year (KWH)			
S.NO	PV Sol	PV Syst	SAM
1	824076.2	83038	80401.23

Fig. 14. 3D image of institute in PV Sol

The Fig. 14 shows the 3D image of the educational institute which is drawn in PV sol. The PV plant efficiency for the different regions in the European countries was analyzed using the PV sol [22]. The author concluded that output from five different regions under study is not same due to the different weather condition. Author proved that there is great dependency on the output of PV plant due to environmental conditions.

5 Conclusion

The detailed study of different simulation tools used for the designing the renewable sources was discussed in this paper. In this paper we clearly discussed each simulation tool and it is found that each tool is having own advantage over other tool. It is clear from the above discussion that Homer Pro is only the tool is having capabilities to design the Hybrid renewable energy sources and this feature is not available in the other tools. Homer Pro optimizes the size of the individual energy sources with respect to constraints defined in the design stages. Homer Pro also gives project owner to select the financially feasible hybrid configuration among the different configuration under study for the project. Solar Pro software optimizes the size of the PV plant based on the available area of the site under study and this software support for optimize the size of inverter based on outcome of the I-V curve characteristics and shadow analysis of the project. PV Syst tool is used to study the subsystem losses in the project and it calculates the performance ratio of the PV plant. Helioscope is used to freeze the PV plant layout which includes the layout of PV module, Inverter and the cable layout. PV Sol is similar to Helioscope only difference between these two tool is PV Sol is having shadow mapping analysis tool due to this output prediction from the PV Sol is accurate than the Helioscope. Finally, it can be conclude that the designer / process owner can select the any of the tool based on their requirements and project needs.

Acknowledgement. The authors are highly thankful to All India Council for Technical Education, New Delhi, India under Research Promotion Scheme, File No. 8–119/FDC/RPS (POLICY-1)/ 2019–20 for their financial grant towards this project.

References

1. Nehrir, M.H., et al.: A review of hybrid renewable/alternative energy systems for electric power generation: configurations, control, and applications. IEEE Trans. Sustain. Energy **2**, 392–403 (2011)
2. Bird, L., Milligan, M., Lew, D.: Integrating variable renewable energy: challenges and solutions. Nat. Renew. Energy Lab. (2013)
3. Ruggiero, S., Lehkonen, H.: Renewable energy growth and the financial performance of electric utilities: a panel data study. J. Clean. Prod. **142**, 3676–3688 (2017)
4. Mekontso, C., et al.: Review of optimization techniques for sizing renewable energy systems. Comput. Eng. Appl. J. **8**(1), 13–30 (2019)
5. Denholm, P., Margolis, R.: Energy storage requirements for achieving 50% solar photovoltaic energy penetration in California. No. NREL/TP-6A20–66595. National Renewable Energy Lab.(NREL), Golden, CO (United States) (2016)
6. Krishna, K.S., Kumar, K.S.: A review on hybrid renewable energy systems. Renew. Sustain. Energy Rev. **52**, 907–916 (2015)

7. Afzal, A., Kumar, H., Sharma, V.K.: Hybrid renewable energy systems for energy security using optimization technique. In: 2011 International Conference & Utility Exhibition on Power and Energy Systems: Issues and Prospects for Asia (ICUE). IEEE (2011)

8. Ram, M., et al.: A comparative analysis of electricity generation costs from renewable, fossil fuel and nuclear sources in G20 countries for the period 2015–2030. J. Cleaner Prod. **199**, 687–704 (2018)

9. Parhizi, S., et al.: State of the art in research on microgrids: a review. IEEE Access **3**, 890–925 (2015)

10. Kumar, P., Deokar, S.: Designing and simulation tools of renewable energy systems: review literature. In: Saeed, K., Chaki, N., Pati, B., Bakshi, S., Mohapatra, D.P. (eds.) Progress in Advanced Computing and Intelligent Engineering. AISC, vol. 563, pp. 315–324. Springer, Singapore (2018). https://doi.org/10.1007/978-981-10-6872-0_29

11. Thula, M., Kumar, M.N., Reddy, V.S.: Simulation and performance analysis of 100kWp solar rooftop using Solar Pro software. In: 2017 Innovations in Power and Advanced Computing Technologies (i-PACT). IEEE (2017)

12. Sharma, S., Kurian, C.P., Paragond, L.S.: Solar PV system design using PVsyst: a case study of an academic institute. In: 2018 International Conference on Control, Power, Communication and Computing Technologies (ICCPCCT). IEEE (2018)

13. Baqir, M., Channi, H.K.: Analysis and design of solar PV system using Pvsyst software. Mater. Today Proc. (2021)

14. Kumar, R., et al.: Design and simulation of standalone solar PV system using PVsyst software: a case study. Mater. Today Proc. **46**, 5322–5328 (2021)

15. Sharma, R., Sharma, S., Tiwari, S.: Design optimization of solar PV water pumping system. Mater. Today Proc. **21**, 1673–1679 (2020)

16. Saraswat, R.: Comparative performance evaluation of solar PV modules from different manufacturers in India by using PVsyst. In: 2016 IEEE 1st International Conference on Power Electronics, Intelligent Control and Energy Systems (ICPEICES). IEEE (2016)

17. Ali, M.S., et al.: Helioscope based design of a MWp solar PV plant on a marshy land of Bangladesh and prediction of plant performance with the variation of tilt angle. GUB J. Sci. Eng. **5**(1), 1–5 (2018)

18. Piyushkumar, P.K., Shinde, S., Shah, P.: Design and estimation of solar PV system for educational campus. Asian J. Converg. Technol. (AJCT) **5**(3), 78–82 (2019). ISSN-2350-1146

19. Khalil, L., et al.: Optimization and designing of hybrid power system using HOMER pro. Mater. Today Proc. **47**, S110–S115 (2021)

20. Sauer, K.J., Roessler, T.: Systematic approaches to ensure correct representation of measured multi-irradiance module performance in PV system energy production forecasting software programs. In: 2012 38th IEEE Photovoltaic Specialists Conference. IEEE (2012)

21. Al Mehadi, A., et al.: Design, simulation and analysis of monofacial solar PV panel based energy system for university residence: a case study. In: IOP Conference Series: Materials Science and Engineering, vol. 1045, no. 1. IOP Publishing (2021)

22. Husu, A.G., et al.: Software Tools for PV Applications in Different Regions of Europe–Part 1: Energy Efficiency

Optimal Power Tracking for Grid-Connected Doubly Fed Induction Generator (DFIG) Wind Turbines Using OPO Algorithm

Samyuktha Penta[1]($\boxtimes$), S. Venkateshwarlu[2], and K. Naga Sujatha[1]

[1] Jawaharlal Nehru Technological University, Hyderabad, Telanagana, India
samyuktha.penta@gmail.com, knagasujatha@jntuh.ac.in
[2] CVR College of Engineering, Hyderabad, Telanagana, India

Abstract. Enhancing Green Energy Development and Mitigating Emissions: A Dual-Focused Approach in Renewable Energy Initiatives The advancement of alternative green energy sources, such as wind energy, and the reduction of greenhouse gas emissions form a two-pronged strategy for renewable energy projects. The integration of power electronic-based controls has enabled Wind Energy Conversion Systems (WECS) to generate a consistent electric power output, irrespective of variations in the wind profile. As one of the most widely utilized renewable sources, wind energy plays a pivotal role in achieving sustainable power generation. This study canters on optimizing Perturb and Observe (P&O) algorithms, presenting a novel solution to address the shortcomings of current methods. Many existing approaches omit the initial tracking phase and assume an incorrect optimal generator speed, overlooking the inertia of WECS. The proposed Optimized Perturb and Observe (OPO) algorithm introduces a swift Maximum Power Point Tracking (MPPT) technique, employing innovative tracking methods to identify the optimal generator speed (Gs) in proximity to the Maximum Power Point (MPP). This enhances the efficiency and reliability of existing P&O algorithms. The research employs three control loops, incorporating Machine Learning (ML) techniques to optimize the P&O control. The primary focus is on analyzing and validating the performance of the proposed OPO algorithm for MPPT control systems. Leveraging the convergence capabilities of the optimization method and the global search capabilities of swarm intelligence, the research aims to maximize power output and contribute to the ongoing efforts in sustainable energy solutions.

Keywords: P & O control · ML algorithm · The MPPT control system

1 Introduction

Power is basic to our day-to-day routines. By and large, the age of electric energy included the ignition of petroleum derivatives, which caused significant issues for people and the planet's current circumstance [1]. Therefore, late endeavors have focused on creating elective techniques for producing power from perfect and economical energy assets. Because of their accessibility and natural cordiality, RESs have as of late been broadly

© The Author(s), under exclusive license to Springer Nature Switzerland AG 2024
S. L. Gundebommu et al. (Eds.): REGS 2023, CCIS 2081, pp. 175–190, 2024.
https://doi.org/10.1007/978-3-031-58607-1_13

utilized [2]. By 2030, one of the most encouraging energy sources, wind energy might give 20% of the world's energy requests [3]. Dissimilar to other energy sources, the breeze is a variable, restricted, non-uniform asset. The breeze is erratic, so factor speed turbines are expected to capitalize on this normal asset [4]. There is an ideal rotational speed for each wind speed since wind turbines might change their rotational speed in light of varieties in the speed of the breeze. The speed that turns out best for every interesting breeze turbine is the ideal Tip Speed Proportion (TSR) [5]. While working safely and proficiently, contemporary breeze turbines' confounded control frameworks should be made due. The three most significant factors that should be taken care of are the edge point, generator force, and matrix side regulator. With the assistance of these administrative methodologies, energy result can be expanded, static and dynamic mechanical anxieties can be decreased, and the lattice can constantly have the energy it needs [6].

Wind Energy Conversion Systems (WECSs) have gone through critical headways because of the significant ascent in modern usage of wind energy [7]. The Breeze Turbine (WT), a focal part of WECS, is sorted into Fixed-Speed Wind Turbines (FSWT) and Variable-Speed Wind Turbines (VSWT) in light of their ability to get through various breeze speeds. Precisely assessing power in an energy change framework presents difficulties, especially because of the moment to-minute vacillations in wind speed, making it hard to guarantee a consistent power yield. To resolve these issues, Wind Energy Transformation Frameworks (WECS) consolidate most extreme power direct following innovation toward answer really to unfavorable atmospheric conditions [9]. Be that as it may, executing this technique to accomplish exceptionally exact and viable results requires a significant venture of time and exertion. The streamlining of following systems depends on deciding the reference or ideal generator speed [10] for motor speed guideline. These methodologies, requiring less framework boundaries, lay out a reasonable association between different WT boundaries, for example, wind speed, generator speed, and mechanical power.

To maximise power extraction, we can use a wide variety of Maximum Power Point Tracking (MPPT) algorithms and methodologies. Each of these models—from the most basic deterministic methods to the most complex predictive ones—has its own set of benefits and drawbacks. As computing power becomes more affordable and accessible, machine learning models may one day supersede many older, more complicated algorithms and methodologies used to determine how much electricity renewable energy sources can provide. Modern advancements in computing have opened up a world of practically endless opportunities for machine learning applications [14]. Because it can learn new things without human intervention, machine learning has quickly become an indispensable tool in many industries. Finding the most efficient way to generate as much power as possible is the primary goal of this research. Developing an accurate tracking device for maximum power locations within Wind Energy Conversion Systems (WECS) is the main purpose of this paper.

2 Background

The maximum power point is calculated by most energy systems nowadays by using Perturb and Observe (P&O). In this conventional strategy [15, 34], a generator is used as a straightforward module. Depending on the load's voltage and current, P&O logic determines the precise power [16, 26]. By endlessly looping the power at any given moment, numerous algorithms use this straightforward logic to return the highest power point. The easiest method to obtain an exact maximum power point is P&O, but this method is sometimes inadequate [17, 35] due to known problems with the technique. In the industry, modified P&O is frequently utilised. Numerous research papers and publications have discussed various MPPT models and methodologies. The bulk of altered P&O networks in WECS are made up of a connected device for generating, distributing, and storing energy, together with a wind turbine, controllers, and other components. Updated MPPT approaches add a new block to the original MPPT process in order to boost power extraction and consequently power [18, 36, 37]. In this research, we aim to provide a unique incremental approach based on machine learning that will enhance the general efficacy of the currently used MPPT method. A machine learning block has been added to this P&O model with the objective of improving the effectiveness of the traditional P&O algorithm-based MPPT approach. In the current way, buck and buck-boost converters directly control how a P&O system is installed [19, 38]. The suggested technique produces an electrical torque proportionate to the square of the rotor speed because it employs three sets of rules that dynamically alter the proportional coefficient in real time. The necessary electrical torque is determined by feedback linearization in the first control law, which assumes that the appropriate rotor speed and power capture coefficient are instantly determined. The second control rule, which is based on a Lyapunov-based analysis [39], estimates the power capture coefficient in real-time, while the third control law establishes the necessary rotor speed. The turbine adjusts its rotor speed adaptively at the operational point, pushing the power in the direction of rising, to reach the desired speed and power capture coefficient.

Several MPPT controllers have gained significant attention, including P&O and/or hill climbing and INC [21–23]. Such approaches, however, are incapable of tracking the stochastic wind behaviour and which considerably deteriorates power quality (PQ). The P&O approach mostly produces inaccurate answers and fluctuations around the MPP, and it also necessitates linking one or more adjustments for widespread use. Although the INC approach can avoid the difficulties described by the P&O method, it does need pretty complex detecting equipment. Furthermore, choosing the step and threshold is difficult [24–26]. As a result, soft computing MPPT solutions are developed to address such issues. Furthermore, one of the key goals of this technique is the simplicity of computations and the lack of the necessity for exact problem information. In comparison to existing approaches, the FLC MPPT method has a quicker tracking speed and fewer variations at the MPP in a variety of conditions [27–29]. Furthermore, unlike the ANFIS [30, 31] and ANN [32, 33], this technique does not require any information training, making it applicable to a wide range of solar systems.

3 Machine Learning–Based Model of Wind Turbines During Maximum Power Point Tracking

Research and development of more efficient power electronics can improve power quality and make wind installations more grid-integrated. As fossil fuel sources continue to dwindle, there is a growing interest in wind energy systems. However, there are a number of obstacles that must be overcome before wind power can reach its full potential. Improving overall productivity requires optimizing power utilization. The Perturb and Observe (P&O) method might not work when the wind speed changes quickly, thus other methods for maximum power tracking are used instead. Regrettably, the traditional method of "perturb and observe" necessitates the introduction of a signal that strays from the trajectory of maximum power points.

Three control loops are used in the suggested strategy to overcome these obstacles. The square of the rotor speed is used to determine the electrical torque, and three control rules allow for real-time adaptive modifications of the proportional coefficient. After the goal electrical torque and rotor speed have been determined, the first control law uses feedback linearization to quickly determine the value. The second control rule establishes the values of the power capture in real time, and the third law is in charge of producing the correct rotor speed. This comprehensive strategy is designed to improve wind energy systems' efficiency, grid integration, and power quality while overcoming the shortcomings of conventional methods (Fig. 1).

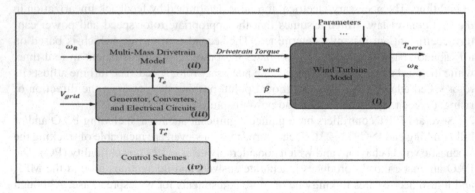

Fig. 1. Block diagram of MPPT control

There are three different places that are fitted with wind turbines, and in order to generate the most amount of electricity possible, a turbine needs to be able to adjust to different wind speeds. In this context, the term "region" refers to the particular locality. In this particular instance, the control mechanism that is responsible for generating torque is responsible for monitoring the speed of the rotor while fixing the pitch angle. In order to maintain a continuous output of rated power, the turbine operates in area III, which is the sector where the amount of wind energy collected is more than what can be measured. However, because there is an abundance of wind energy in this region, efficiency is at a significant disadvantage. Pitch control is used to limit rotor speed, and the generating torque is maintained at a constant level. This helps to reduce fluctuations in output power. Given that the output power is affected by the product of rotor speed and torque, it is of the utmost importance to keep it at the value that is specified for it. In Region III, it is possible to implement strategies that involve constant torque as well as constant power.

Blade pitch and generator torque work together to control the speed of the rotor in Region II, which is located between Regions I and III. Due to the fact that it is situated in the middle, area II is sometimes disregarded, despite the fact that it is connected to the area representing the most power. It is possible to effectively manage the rotor speed in this zone thanks to the complicated interplay between the blade pitch and the generator torque, which in turn optimises the overall performance of the turbine (Fig. 2).

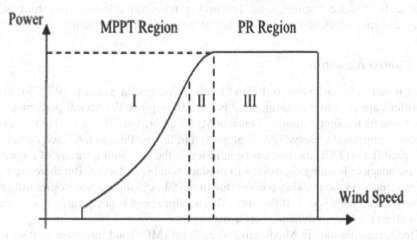

Fig. 2. Operating regions of wind turbine

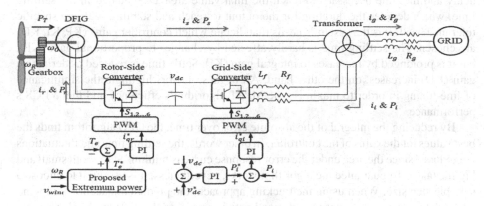

Fig. 3. Proposed MPPT Control

In Fig. 3, we can see the stator circuit in action, with rotor circuits connected via slip rings and 3-phase converters. For wind turbines that produce power in the megawatt range, the Doubly Fed Induction Generator (DFIG) is now the go-to option. The aero-dynamic system needs to be able to work well throughout a wide variety of wind speeds in order to determine the optimal tip-speed ratio and optimise aerodynamic perfor-mance. As a result, DFIG systems are able to function in both sub-synchronous and super-synchronous phases because to rotor speeds that are almost synchronous.

An AC-DC-AC converter is built into the induction machine rotor circuit. The sole power that needs converter approval is the electronically provided power to the rotor, which is usually about 30% of the nominal generator output. When compared to config-urations where the power electronic converter handles all of the power, this partial power rating lowers the system cost. Furthermore, this method reduces converter losses to a minimum. We will explore the basic components and typical operations of DFIG devices in the context of wind power applications, using the traditional induction generator as an example. Power converters, related control systems, variable-speed capabilities, and issues relevant to each application will be the focus of our investigation.

3.1 Control Algorithm

When it comes to variable-speed Wind Energy Conversion Systems (WECS), MPPT algorithms are vital for capturing wind power. To improve the overall performance of WECS and fix the shortcomings of current MPPT algorithms, it is crucial to introduce a suitable Perturb and Observe (P&O) algorithm in this section, as indicated earlier. The Tip-Speed Ratio (TSR) method has been tested in the past with a variety of algorithms that use adaptive learning controllers to produce wind speed input. But there are major inconsistencies in the tracking process due to mistakes in the methodologies utilised to measure actual wind speed. If the optimal generator speed is not accurately calculated, it can affect the overall dynamic performance.

Reduce oscillations in Maximum Power Point (MPP) and minimise settling times with the use of the Optimal Perturb and Observe (OPO) method. Both the propagation delay and the time necessary to reach the final value area are included in the settling time, which denotes the duration for the output to reach and stabilise within a specific tolerance band. It is the system's needs that dictate which controller gains (KP, KI, KD) are chosen. The settling time is hardly affected by changes to proportional gain (KP), but it is prolonged by increases to integral gain (KI). Settle time is reduced as derivative gain (KD) increases, on the other hand. All of these factors highlight the significance of fine-tuning in order to maximise the MPPT algorithm's efficiency and the WECS's performance.

By reducing the integral of the absolute error over time, the OPO algorithm finds the best values for the gains of the controller. In other words, the settling time and fluctuations are reduced since the area under the error response curve is minimised for both small and big mistakes. To guarantee the algorithm works, nevertheless, it is essential to choose a suitable step size. When using the tracking approach, the predicted inaccuracy in wind speed data is ignored. Precise wind speed data is crucial to the TSR method's tracking strategy.

Then again, the OPO calculation exploits the projected breeze speed. Regardless of their straightforwardness and effectiveness, the TSR and OPO calculations don't consider the mistake in estimating wind speed that influences generator speed and mechanical power. A reasonable step size should likewise be chosen for the OPO calculation's following stage. A few current strategies overlook the primary following stage and expect that the producing recurrence works at a wrong ideal generator speed that ignores WECS dormancy. It matches the prior limitations and is risky in bigger organizations. The recommended OPO calculation utilizes creative following techniques in view of recognizing the best made up position of generator speed G_s near the MPP to work on the effectiveness and steadfastness of the current P&O calculations. Involving strategies for anticipating wind speed, the anticipated ideal generator not entirely set in stone and is expected to falsely run. Beginning from this phony spot, it then searches for the genuine area of the generator's greatest speed. Thus, a similar condition utilized in the OPO approach is applied to decide the G_s utilizing the projected breeze speed (Fig. 4).

$$u(t) = Kp, e(t) + Ki \int_{o}^{t} e(t)dt + Kd\left(\frac{de(t)}{dt}\right) \tag{1}$$

$$G_s = M \cdot V_{wind} \tag{2}$$

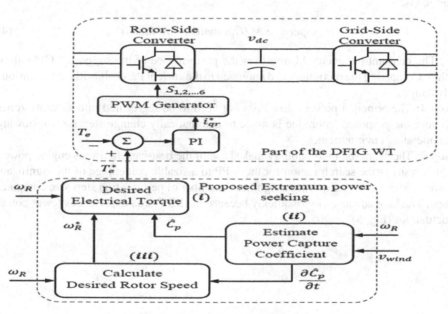

Fig. 4. Rotor side converter control

The wind speed in m/s is represented by Vwind, where M denotes a constant system parameter. To reduce dependence on the system parameter (M), the rated wind and generator speeds are calculated per unit:

$$G_s = Opt.G_s \frac{V_{wind}}{RatedV_{wind}} = \frac{V_{wind}}{12} \qquad (3)$$

In this instance, the generator speed per unit is Opt. G s, with a wind speed of 12 m/s as the maximum. By specifying the G $_s$, the feature selection for the MPP is made smaller, and the necessity for system components and the well-known fault in the Wind Speed Estimation (WSE) technique used is also removed. The following are the key elements of the tracking approach for the proposed methodology:

Step 1: Calculate the Gs and estimate wind speed.

Step 2: For rapidly changing the speed of the generator in close vicinity to the ideal hypothetical power, significant adaptive step-sizes are utilised under G_s.

Step 3: Otherwise, MPP operation with minimal power fluctuations is ensured using the OPO algorithm and the adjustable step-size S. The needed step sizes and the operating speed that is achieved by combining the quick tracking approach are calculated using the following equation. By analysing the difference between the real and ideal generator speeds, one can see that these step sizes start big and then steadily get smaller until they hit the Gs.

$$\Delta Speed = M \, (OptimalGs - ActualGs) \qquad (4)$$

The constant parameter M improves the performance of the suggested OPO algorithm. Comparing the recommended approach to the current P&O algorithms has various advantages.

Step 4: The principal power curve does not need to be segregated into separate zones because the proposed technique is made to automatically change step sizes following the operating environment.

Step 5: The proposed methodology makes use of the best-case fictitious engine power, which restricts the search region for the MPP to a sizable percentage of the significant power curves without requiring a significant number of perturbation step-size computations. Tracking advances more quickly because the MPP area experiences fewer power fluctuations (Fig. 5).

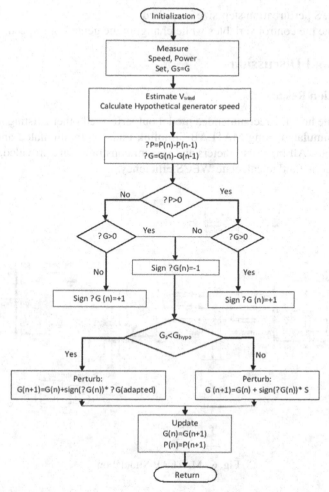

Fig. 5. Optimized P&O Algorithm

The proposed method is highlighted because of its simple tracking strategy. It also corrects the flaw in the method currently being used to measure wind speed in a variety of environmental circumstances. You can better comprehend the suggested algorithm's tracking method by looking at the flowchart below.

Step 1: Calculate the real generator speed and optimal mechanical power, and wind speed.
Step 2: Calculate the ideal generator speed and power difference.
Step 3: Indicate the direction of the disturbance.
Step 4: Indicate the perturbation step-sign Size's
Step 5: Compare the speeds of the generators.
Step 6: Declare and determine the ideal perturbation step size.
Step 7: Speed is applied below the Gs.

Step 8: If not, S perturbation step-size is applied.
Step 9: Update the control variables while changing the generator's speed.

4   Result and Discussion

4.1   Simulation Results

To demonstrate how the recommended model outperformed other existing MPPT algorithms, it is simulated using MATLAB/Simulink under both simulated and real wind speed variations. All input parameters and system constraints are provided, along with a conventional method to calculate WECS efficiency.

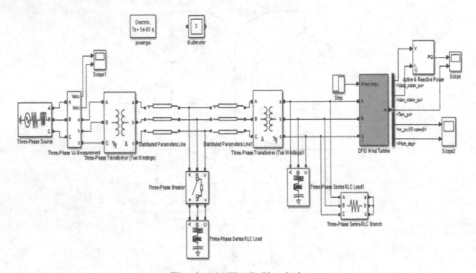

Fig. 6. MATLAB Simulation

This section models the grid-connected Wind system using MATLAB software. This system, which consists of power electrical components and controllers, is depicted in Fig. 6.

This article discusses a DPC-based multiple-target implementation technique for achieving the three control targets for DFIG under distorted and unbalanced grids: balanced and sinusoidal stator current, smooth stator active and reactive powers, and smooth electromagnetic torque and stator reactive power. The outcomes of the trial validated the value of the suggested control strategy. The suggested DPC technique's key benefit over traditional Voltage Oriented Control (VOC) and Direct Power Control (DPC) methods is that it offers a quicker dynamic reaction and a more accurate steady state without requiring a laborious grid voltage series breakdown or reference computing. The proposed method is unaffected by changes in the DFIG parameter and employs a vector proportional–integral (VPI) regulator set at resonance frequencies of 100 and 300 Hz. Whenever control objective II of smooth active and reactive powers is accomplished, the third harmonic component of the stator current manifests, but the negative component is suppressed. As control objective III is completed, the electromagnetic torque and the pulses of reactive power at 100 and 300 Hz from the stator both decrease (Fig. 7).

4.2   Comparative Analysis

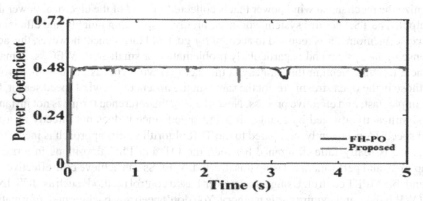

Fig. 7. Comparison of the power coefficient of the proposed method

In spite of rapid variations, the FH-PO algorithm maximises power generation while minimising power loss with respect to the optimal speed of the generator [20]. At last, the solution that has been described can deal with rapid fluctuations in wind speed effectively and efficiently.

Accurately tracking the MPP is a major issue for the MPPT algorithms because to the random fluctuations in wind speed. With a 20% turbulence intensity ratio and an average wind speed of 8 m/sec, the proposed method is tested for resilience in the face of sudden and unpredictable changes in wind speed. Figure 8 shows that when comparing the two methods, the suggested IR-PO algorithm achieves better results in keeping an eye on the optimal Cp and tip-speed-ratio values. The speed of the generator and the mechanical power that is collected are both affected by this

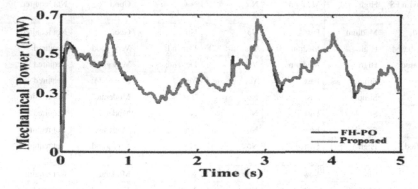

Fig. 8. Comparison of mechanical power of the proposed method

Several conventional and ML-enabled MPPT methods are compared in Table 1 with respect to complexity, convergence speed, performance, memory requirement, and previous training. A variable speed wind turbine's optimum power point tracking is the main objective of the maximum power point tracking (MPPT) algorithm. Choosing the optimal MPPT method could be challenging. Even though they are less efficient, algorithms like TSR OT and PSF, which are based on indirect power control and are simple and fast, optimize the mechanical wind power that is collected instead of the electrical power that is output. The TSR control system works admirably, responding quickly and efficiently. A precise anemometer is required to account for gust and turbulence; however, this adds expense to the system and is particularly problematic for small-scale WECSs. A major obstacle to implementing this strategy is the fact that wind speeds close to the turbine and those in the open stream are not the same. In the absence of a wind speed sensor, OT is a simple, fast, and effective process. Nevertheless, the reference torque is not instantly and significantly affected by changes in wind speed since it does not directly measure wind speed. Consequently, compared to the TSR algorithm, this approach is inefficient.

There is really little difference between the OT and PSF algorithms in terms of complexity and performance. Using this method, WECSs can achieve cost-effective and dependable MPPT control. Using direct power-based control methods such as HCS, INC, and ORB is easy and requires little memory. You don't need any background information or wind speed sensors to use these approaches to find the optimal electrical power. Unfortunately, the algorithms' lacklustre performance when faced with fluctuating wind conditions limits their applicability in some scenarios. In addition to being inexpensive and dependable, these algorithms are sensor less. The HCS method is popular and simple to use since it does not necessitate the measurement of mechanical factors such as rotor speed, turbine speed, or wind speed. The algorithm's tracking performance is unaffected by changes to the parameters of the turbine or generator since it is not system

Table 1. Comparison of various MPPT algorithm based on different characteristics

Algorithm	Complexity	Convergence speed	Memory requirement	Wind speed measurement	Performance under varying wind conditions	Prior training/knowledge
HCS with FS & AS	High	Medium	No	No	Good	Not required
Hybrid	Medium	Fast	No	No	Good	Not required
Fuzzy-based	High	Medium	Yes	Depends	Very good	Required
NN-based	High	Medium	Yes	Depends	Very good	Required
PSF	Simple	Fast	Yes	Yes	Moderate	Required
TSR	Simple	Fast	No	Yes	Moderate	Not required
OT	Simple	Fast	No	No	Moderate	Required
HCS	Simple	Low	No	No	Moderate	Not required
Modified HCS	High	Fast	No	No	Very good	Not required
INC	Simple	Low	No	No	Moderate	Not required
Proposed OPO	Medium	Fast	Yes	Yes	Very good	Required

dependent. The maximum power of every wind velocity can be obtained using the HCS approach. But getting to Maximum Power Point (MPP) is a drag and tracking drains power significantly.

The outcome can be the stalling of smaller wind turbines. The updated HCS algorithm fixes the weaknesses of the original, which were a lack of responsiveness and an incorrect directionality when the wind was blowing at high speeds. Additionally, the HCS slows tracking and produces hunting (oscillation) around the MPP if the step size is too tiny. When dealing with problems where the step size is excessively large, the fixed and adaptive step HCS approach can be helpful. No additional sensors are needed to detect wind and rotor speed by either the HCS or INC methods. The INC method has the ability to provide more efficient power tracking of MPP compared to the HCS algorithm. When it comes to the wind system, both methods are simple and flexible, although variations near the MPP reduce its efficiency. Increased MPPT performance, convergence speed, and system precision are achieved by automatically modifying the step size while following the maximum power point of the wind power system using the upgraded INC algorithm. The ORB MPPT method is easy to understand and implement since it only requires two measurements: dc voltage and dc current. Due to the fact that mechanical sensors and prior knowledge of the energy system are not required, this technology is simple, autonomous, and customisable. Also, it does a good job of tracking wind energy. If an algorithm has flaws, they can be fixed by modifying it or by employing a hybrid approach. Adaptive algorithms, soft computing-based algorithms based on fuzzy logic and neural networks, and other maximum power point tracking (MPPT) methods require knowledge of the system in advance, but can efficiently control nonlinearity and predict the ideal power. The complexity of the system determines the optimal number of controller rules, which in turn determines the efficacy of fuzzy control-based methods. With NN-based MPPT control, the system's dynamic speed and power responses are better balanced. In most cases, NN control systems work well since mechanical components tend to change with age and exposure to different environments. Regular training of the neural network is necessary to provide accurate MPPT. When coping with unpredictable fluctuations in wind speed and power demand, the OPO MPPT algorithm excels above other algorithms due to its greater adaptability, reliability, and accuracy.

5 Conclusion

High convergence velocity, improved search performance, and an easier implementation process are just a few of the notable benefits of the proposed Optimum Perturb and Observe (OPO) algorithm for Maximum Power Point Tracking (MPPT), an advanced method for optimizing power extraction from wind turbines. The main goal of this technique is to achieve the optimal power capture coefficient adaptively, regardless of how fast the wind is changing. In order to confirm its capability for producing high-power, the algorithm-based wind-driven power system is painstakingly modelled in MATLAB and subjected to extensive testing. The OPO algorithm, in contrast to conventional approaches, does not depend on perturbation signals. Further to its adaptability and resilience, it does not presume any prior information regarding the optimal tip-speed ratio and maximum power capture coefficient. A very efficient and successful approach is produced by merging the optimization method's convergence capability with swarm intelligence's global search capability. In terms of reaching Maximum Power Point (MPP) and tracking efficiency, the experimental results confirm that the OPO-based MPPT algorithm outperforms both the standard and advanced MPPT methods. Its ability to improve the overall performance of wind energy systems is highlighted by its faster and superior convergence.

References

1. Mohtasham, J.: Renewable energies. Energy Procedia **74**, 1289–1297 (2015)
2. Ganiyu, S.O., Martinez-Huitle, C.A., Rodrigo, M.A.: Renewable energies driven electrochemical wastewater/soil decontamination technologies: a critical review of fundamental concepts and applications. Appl. Catal. B Environ. **270**, 118857 (2020)
3. Darwish, A.S., Al-Dabbagh, R.: Wind energy state of the art: present and future technology advancements. Renew. Energy Environ. Sustain. **5**, 7 (2020)
4. Chavero-Navarrete, E., et al.: Expert control systems for maximum power point tracking in a wind turbine with PMSG: state of the art. Appl. Sci. **9**(12), 2469 (2019)
5. Golnary, F., Moradi, H.: Dynamic modelling and design of various robust sliding mode controls for the wind turbine with estimation of wind speed. Appl. Math. Model. **65**, 566–585 (2019)
6. Geertsma, R.D., et al.: Design and control of hybrid power and propulsion systems for smart ships: a review of developments. Appl. Energy **194**, 30–54 (2017)
7. Mojumdar, M.R.R., et al.: Electric machines & their comparative study for wind energy conversion systems (WECSs). J. Clean Energy Technol. **4**(4), 290–294 (2016)
8. Tahiliani, S., Sreeni, S., Moorthy, C.B.: A multilayer perceptron approach to track maximum power in wind power generation systems. In: TENCON 2019–2019 IEEE Region 10 Conference (TENCON), pp. 587–591. IEEE (2019)
9. Naidu, R.P.K., et al.: A review on PMSG based wind energy conversion system. J. Algebr. Stat. **13**(3), 1058–1065 (2022)
10. Mousa, H.H., et al.: Performance assessment of robust P&O algorithm using optimal hypothetical position of generator speed. IEEE Access **9**, 30469–30485 (2021)
11. Dileep, G.: A survey on smart grid technologies and applications. Renew. Energy **146**, 2589–2625 (2020)
12. Kumar, D., Chatterjee, K.: A review of conventional and advanced MPPT algorithms for wind energy systems. Renew. Sustain. Energy Rev. **55**, 957–970 (2016)
13. Esram, T., Chapman, P.L.: Comparison of photovoltaic array maximum power point tracking techniques. IEEE Trans. Energy Convers. **22**(2), 439–449 (2007)
14. Stetco, A., et al.: Machine learning methods for wind turbine condition monitoring: a review. Renew. Energy **133**, 620–635 (2019)
15. Linus, R.M., Damodharan, P.: Maximum power point tracking method using a modified perturb and observe algorithm for grid connected wind energy conversion systems. IET Renew. Power Gener. **9**(6), 682–689 (2015)
16. Soetedjo, A., Lomi, A., Mulayanto, W.P.: Modeling of wind energy system with MPPT control. In: Proceedings of the 2011 International Conference on Electrical Engineering and Informatics, pp. 1–6. IEEE (2011)
17. Lodhi, E., et al.: Performance analysis of 'Perturb and Observe' and 'Incremental Conductance' MPPT algorithms for PV system. In: IOP Conference Series: Materials Science and Engineering, vol. 220, no. 1, p. 012029. IOP Publishing (2017)

18. Ferdous, S.M., et al.: Design and simulation of an open voltage algorithm based maximum power point tracker for battery charging PV system. In: 2012 7th International Conference on Electrical and Computer Engineering. IEEE (2012)
19. Atallah, A.M., Abdelaziz, A.Y., Jumaah, R.S.: Implementation of perturb and observe MPPT of PV system with direct control method using buck and buck-boost converters. Emerg. Trends Electr. Electron. Instrum. Eng. Int. J. 1(1), 31–44 (2014)
20. Ali, A.I., Sayed, M.A., Mohamed, E.E.: Modified efficient perturb and observe maximum power point tracking technique for grid-tied PV system. Int. J. Electr. Power Energy Syst. **99**, 192–202 (2018)
21. Mousa, H.H., Youssef, A.R., Mohamed, E.E.: State of the art perturb and observe MPPT algorithms based wind energy conversion systems: a technology review. Int. J. Electr. Power Energy Syst. **126**, 106598 (2021)
22. Lalouni, S., et al.: Maximum power point tracking based hybrid hill-climb search method applied to wind energy conversion system. Electr. Power Compon. Syst. **43**(8–10), 1028–1038 (2015)
23. Cheng, W.: Incremental conductance algorithm for maximum wind power extraction using permanent magnet synchronous generator. Diss. (2013)
24. Li, J., Wu, Y., Ma, S., Chen, M., Zhang, B., Jiang, B.: Analysis of photovoltaic array maximum power point tracking under uniform environment and partial shading condition: a review. Energy Rep. **8**, 13235–13252 (2022)
25. Bhukya, L., Kedika, N.R., Salkuti, S.R.: Enhanced maximum power point techniques for solar photovoltaic system under uniform insolation and partial shading conditions: a review. Algorithms **15**(10), 365 (2022)
26. Sarwar, S., et al.: A novel hybrid MPPT technique to maximize power harvesting from PV systems under partial and complex partial shading. Appl. Sci. **12**(2), 587 (2022)
27. Singh, J., Singh, S.P., Verma, K.S., Kumar, B.: Comparative analysis of MPPT control techniques to enhance solar energy utilization and convergence time under varying meteorological conditions and loads. Front. Energy Res. **10**, 856702 (2022)
28. Louarem, S., Kebbab, F.Z., Salhi, H., Nouri, H.: A comparative study of maximum power point tracking techniques for a photovoltaic grid-connected system. Electr. Eng. Electromechanics **4**, 27–33 (2022)
29. George, S., Sehgal, N., Rana, K.P.S., Kumar, V.: A comprehensive review on modelling and maximum power point tracking of PEMFC. Cleaner Energy Syst. **3**, 100031 (2022)
30. Guerra, M.I., de Araújo, F.M., de Carvalho Neto, J.T., Vieira, R.G.: Survey on adaptative neural fuzzy inference system (ANFIS) architecture applied to photovoltaic systems. Energy Sys. 1–37 (2022)
31. Revathy, S.R., et al.: Design and analysis of ANFIS–based MPPT method for solar photovoltaic applications. Int. J. Photoenergy **2022** (2022)
32. Kiran, S.R., Basha, C.H., Singh, V.P., Dhanamjayulu, C., Prusty, B.R., Khan, B.: Reduced simulative performance analysis of variable step size ANN based MPPT techniques for partially shaded solar PV systems. IEEE Access **10**, 48875–48889 (2022)
33. Ali, A.H., Najafi, A.: Optimization and performance improvement of grid-connected PV plant based on ANN-PSO and P&O algorithms. Int. Trans. Electr. Energy Syst. **2022** (2022)
34. Pathak, P.K., Padmanaban, S., Yadav, A.K., Alvi, P.A., Khan, B.: Modified incremental conductance MPPT algorithm for SPV-based grid-tied and stand-alone systems. IET Gener. Transm. Distrib. **16**(4), 776–791 (2022)
35. Jamiati, M.: Modelling of maximum solar power tracking by genetic algorithm method. Iranica J. Energy Environ. **12**(2), 118–124 (2021)
36. Gouabi, H., Hazzab, A., Habbab, M., Rezkallah, M., Chandra, A.: Experimental implementation of a novel scheduling algorithm for adaptive and modified P&O MPPT controller using fuzzy logic for WECS. Int. J. Adapt. Control Signal Process. **35**(9), 1732–1753 (2021)

37. Khan, M.J., Mathew, L.: Artificial neural network-based maximum power point tracking controller for a real-time hybrid renewable energy system. Soft. Comput. **25**(8), 6557–6575 (2021)
38. Bhan, V., et al.: Performance evaluation of perturb and observe algorithm for MPPT with buck-boost charge controller in photovoltaic systems. J. Control Autom. Electr. Syst. **32**(6), 1652–1662 (2021)
39. De Jesus Teran-Gonzalez, R.A., Perez, J., Beristain, J.A.: Nonlinear observer based on linear matrix inequalities for sensorless grid-tied single-stage photovoltaic system. Adv. Electr. Comput. Eng. **21**(3), 91–98 (2021)

Communications and Signal Processing

Image Inpainting for Object Removal Application using Improved Patch Priority and Exemplar Patch Selection

B. Janardhana Rao[1](✉) (iD), K. Revathi[2] (iD), Yalamanchili Bhanusree[3] (iD),
Venkata Krishna Odugu[1] (iD), and Harish Babu Gade[1] (iD)

[1] CVR College of Engineering, Hyderabad, India
janardhan.bitra@gmail.com
[2] Sphoorthy Engineering College, Hyderabad, India
[3] Vallurupalli Nageswara Rao Vignana Jyothi Institute of Engineering and Technology,
Hyderabad, India

Abstract. Image inpainting is a method that can be employed to repair damaged images and remove distracting elements. The effectiveness of image inpainting approach heavily relies on the computation of patch priority and the selection of exemplar patches in exemplar-based methods. The occurrence of the dropping effect in the computation of the most significant patch priority and the occurrence of matching errors in the selection of the best patch are the primary concerns in example inpaint approaches. The upgraded priority calculation approach is utilized to prevent the dropping effect and introduces a new similarity evaluating procedure called Square of Mean Difference (SMD). The effectiveness of the suggested strategies is evaluated by qualitatively evaluating them with the existing methods. The results demonstrate that the suggested methods surpassed the performance of the existing strategies.

Keywords: Matching Error · SMD · Image Inpainting · Patch Priority · Exemplar patch

1 Introduction

Image inpainting is a typical approach for restoring damaged or scratched areas of images. Parts that have been damaged and areas lacking data or color information are filled in during this phase. Image inpainting has a wide range of possible uses, including restoring damaged or outdated images, erasing undesired elements, and filling in blank spaces. There are a few broad types of image inpainting techniques, each with its own set of pros and cons, and they all aim to restore damaged areas in line with specific expectations for the restored image's quality.

Inpainting using Partial differential equations (PDEs): Object removal is the primary goal of PDE-based image inpainting algorithms. Assuming a tiny missing area, their output is acceptable. PDE-based image inpainting techniques are extremely slow and

© The Author(s), under exclusive license to Springer Nature Switzerland AG 2024
S. L. Gundebommu et al. (Eds.): REGS 2023, CCIS 2081, pp. 193–204, 2024.
https://doi.org/10.1007/978-3-031-58607-1_14

produce a fuzzy reconstruction when dealing with huge missing areas. Inpainting using texture synthesis: In order to repair damaged pixels, algorithms that rely on texture synthesis look for nearby pixels that are virtually identical. To cover up the blanks, most of the first inpainting methods copied pixels from nearby areas using texture generation. Only images with textures can benefit from this form of inpainting.

Exemplar based inpainting: Using this inpainting technique yields satisfactory results regardless of the magnitude of the missing region.

Hybrid based inpainting: Reconstructing the damaged image using this method, which combines texture synthesis with PDE-based inpainting techniques, nevertheless has the drawback of producing a fuzzy recovered image when the removed area is not tiny.

2 Literature Review

The isophote direction-based dented region can be effectively and efficiently served by using PDE-based techniques to solve partial differential equations. The regularization term was used by Rathish et al. [1] in their work by taking the square of the L2-norm of the image's Hessian. On the Fourier domain, they solved the resulting semi-discrete approach using convexity separation techniques. Models for image reconstruction based on fractional-order diffusion were suggested by Sridevi and Kumar [2–5]. When used to fill in missing areas and eliminate noise in other areas, these models perform admirably. Models based on fractional calculus [6, 7] are effective tools for enhancing picture quality.

The notion of exemplar-based image inpainting approach was initially proposed by Criminisi et al. in 2003 [8]. The key aspect of this method involves a sampling procedure of the image that is driven by isophotes. In 2004, Criminisi et al. released an expanded version of their prior research in [8], providing a more comprehensive explanation of the method and conducting numerous experiments [9]. In their study, Wen-Huang et al. pointed out the drawbacks of the earlier methods mentioned in references [8, 9], where the priority function defined resulted in a rapid decrease of the confidence term to zero [10]. The aforementioned strategies are limited in their effectiveness in edge zones. Put simply, they are unable to effectively fill a target patch situated in the peripheral areas. In 2010, Zongben et al. proposed a new exemplar-type image inpainting technique that utilizes patch sparsity to produce a high-quality image [11].

In 2014, Jian Wang et al., introduced a novel technique that incorporates both the mean sum of squared difference and normalized cross-correlation, rather than solely relying on the SSD. This algorithm also employs the same priority function as described in a previous study [10]. In their study, H. Chinmayee et al. introduced a novel approach to enhance the rebuilt image visually using exemplar-type inpainting [12]. In this work, the methodology focuses on enhancing the picture inpainting method specifically for images that contain edge and corner information. The data term is determined using fractional derivative and various curvature-finding algorithms in patch priority.

The most similar exemplar patch are determined by Liu et al. [13] using the SSIM. In order to accomplish this, the optimal candidate patch was generated by taking into account four different rotation and inversion scenarios. The golden section was used by Nan et al. [14] to give the data item and the confidence item different weights. The restoration order's reasonableness was the target of this strategy. Nevertheless, it failed to effectively prevent mismatch mistakes, suggesting that the restoration effect could use some work. Yao [15] found that while calculating priority, it was helpful to take into account how similar the target patch was to the surrounding patch. They also changed it such that addition is used instead of multiplication. Janardhana Rao et al. [16] presented a better method for priority computation that incorporates adaptive coefficients and regularization factors. Revathi et al. [17] developed an inpainting method with adjustable patch sizes. The efficiency of the method is evaluated for various patch sizes and concluded which patch sizes taken edge of the target region generating the efficient results. Author [18], discussed the evaluation of exemplar based inpainting methods implemented by various authors in the literature. To determine the example patch, Zhang et al. [19] used a similarity metric that combined mean squared difference and square of mean differences. To inpaint the missing areas produced by removing undesired objects, scratches, compressing the images, or Internet-based image alteration, Abdulla and Ahmed [20] developed an exemplar-based method, one of the utmost essential and prevalent image inpainting approaches. In order to find the data that is a good fit, the suggested technique uses two stages of searching. Initially, the search algorithm scans the entire image for patches that are visually similar and then utilizes the Euclidean distance to pick the ones that are most similar. In the second stage, we find the distance between the chosen patches and the empty patch. Zahra et al. [21], employed adaptive spline interpolation for the purpose of image inpainting. This approach takes into account varying numbers of neighboring pixels in 4-directions for each pixel in the missing block. Ahmed and Abdulla [22] presented an inpainting technique that incorporates a novel approach for efficiently locating the most analogous patch.

Furthermore, exemplar-based approaches to video inpainting have also advanced in recent years [23–27]. Finding the most important patch-enhanced approach to prevent the dropping effect and a new selection process for the similar patches that produces good inpainting results were the goals of this work. These exemplar-based procedures have attracted a lot of interest from scholars, who have been proposing new and improved techniques all the time.

The rest of the sections of the paper are outlined as follows: The issues that have been observed in existing image inpainting techniques outlined in Sect. 2. Section 3 outlines the suggested structure for image inpainting. The outcomes of the experiment and the accompanying arguments are detailed in Sect. 4. The paper ends in Sect. 5.

3 Existing Image Inpainting Using Exemplar Based Method

Two basic operations are executed during the inpainting process in exemplar-based image inpainting approaches. The first step of the work is to find the most significant patch on the outside of the target area. After that, we look for the best patch in the source area that matches the criteria of the most significant patch in the target area. This technique is shown in detail in Fig. 1.

Criminisi et al. [9] first presented this method in their pioneering study. The part need to be inpainted in the image is called the target region (Ω), while the remaining portion is called the source region (Φ). The pixels on the boundary were used as the center pixels ($\partial\Omega$) for each of the patches that were created along the edge of the target area ($\partial\Omega$). After generating a priority function, we were able to find the patch on the target portion boundary that had the highest priority. We found out which patch, with its center at pixel p, was most important.

$$P_p = C_p * D_p \tag{1}$$

where, C_p denotes the confidence term and D_p represents the data term.

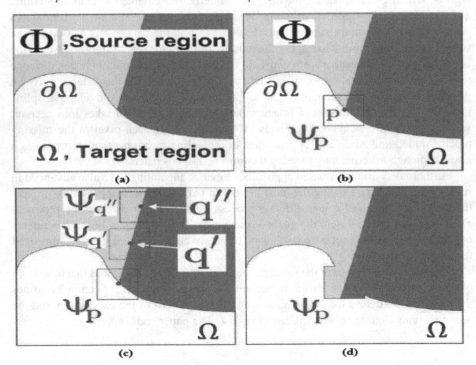

Fig. 1. The basic approach of the exemplar type inpainting method

Then, the best matching patches $(\Psi_{q'}, \Psi_{q''})$ from the source region were placed into the most important patch at the edge of the Ω. A patch that is the most similar to the patches that were produced in the source region may be identified using the SSD distance metric. This is accomplished by evaluating the degree of dissimilarity that exists between the most notable patch and the patches that are included inside the source region. The source region's most desirable candidate for the top-priority patch to fill is the one with the smallest SSD. Equation (2) was used to do this evaluation.

$$\Psi_{q'} = \arg \min_{\Psi_{q^i} \Phi} d_{SSD}\left(\Psi_p, \Psi_{q^i}\right) \tag{2}$$

where,

$$d_{SSD}\left(\Psi_p, \Psi_{q^i}\right) = \sum\left[\left(R_{\Psi_p} - R_{\Psi_{q^i}}\right)^2 + \left(G_{\Psi_p} - G_{\Psi_{q^i}}\right)^2 + \left(B_{\Psi_p} - B_{\Psi_{q^i}}\right)^2\right] \tag{3}$$

in the target region, the red, green, and blue planes are denoted as R_{Ψ_p}, G_{Ψ_p} and B_{Ψ_p}, respectively. Planes $R_{\Psi_{q^i}}$, $G_{\Psi_{q^i}}$ and $B_{\Psi_{q^i}}$ in the source region are red, green, and blue, respectively.

Afterwards, the best example patch $\Psi_{q'}$ was placed into the selected patch $\left(\Psi_p\right)$. The plan was to update the target zone's boundaries and repeat the process until the target area in the picture was finished.

4 Notations

Through the use of an exemplar-based picture inpainting approach, the focus of this study is on the removal of objects and the subsequent filling of the hole that is left behind after it has been removed. An illustration of the parameters that are represented by this method may be seen in Fig. 2. In the image, the source region is denoted by Φ, while the target region is the region that is generated after the undesired object in the image is removed in the process. A boundary of Ω is denoted by the symbol $\partial\Omega$. The patches taken on the edge of the target region are represented as Ψ_{p1}, Ψ_{p2}, Ψ_{p3} and Ψ_{p4} which are shown in Fig. 2, like this many patches are made as pixels on the edge as center pixel. The patches created over the source region are denoted as Ψ_{q1}, Ψ_{q2} and Ψ_{q3}, and shown in Fig. 3, many patches are created on the source region to check the most similar patch.

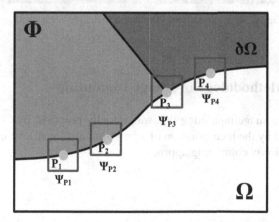

Fig. 2. Identification of patches on the target area or region and corresponding notations.

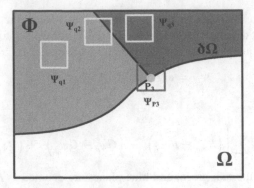

Fig. 3. The patches in the source region and corresponding notations.

The most significant patch is located, and then the most similar patch from the source area is chosen to fill it. For optimal inpainting results, use a similarity metric that fits your needs. The target patch can end up containing unidentified data because of a matching error between the patches in the source region and the most significant patches. The result may be seen in Fig. 4. Consequently, it is crucial to fill the target patch using the appropriate similarity metric.

Fig. 4. Representation of the patch matching error results.

5   Proposed Methodology of Image Inpainting

The exemplar-type image inpainting that was initially proposed by Criminisi et al. [9] has been enhanced by the incorporation of a new similarity verification process as well as an improved priority computing approach.

5.1 Patch Priority Calculation

The present exemplar based inpainting technique fills in the target region with the best patches from the source area. Here, while inpainting, we consider the data about the image's structure and texture. Problems arise with the dropping effect while using this inpainting technique. Because of this impact, the supplied images cannot be accurately inpainted [9]. A better priority calculation method that incorporates an additional parameter S_p and substitutes addition for multiplication of the confidence and data terms successfully resolves the previously noted issue. Equation (4) can be used to determine the target patch's priority.

$$P_p = a * C_p + b * D_p + c * S_p \tag{4}$$

The structure, texture and linear structure constants in the target region are denoted by a, b, and c, respectively. The data available in the source location is utilized to ascertain the pertinent constant value, which is found to be elevated.

The amount of reliable data available near the p pixel is represented by the term C_p, which is also referred to as the confidence term. The subsequent equation was employed in the computation:

$$C_p = \frac{\sum_{t \in (\Psi_p \cap \Phi)} C(t)}{|\Psi_p|} \tag{5}$$

where the size of the target patch in pixels is denoted by $|\Psi_p|$ and t stands for the coordinates of its pixels.

We take into account the variable D_p, which represents the amount of isophotes that cross the boundary at pixel p.

$$D_p = \frac{\left|\nabla I_p^{\perp} \cdot n_P\right| / L_p + d}{255} \tag{6}$$

The geometric qualities of the isophote, namely its curvature along the isophote's direction, are represented by the variable L_p. Information can be more easily propagated towards the isophote with the addition of curvature. It is thought of as curvature.

$$L_p = \nabla \cdot \left(\frac{\nabla I_p}{|\nabla I_p|}\right) \tag{7}$$

The term S_p denotes the specific arrangement of the measurement function at a local level. The computation is carried out using the ideas of structural tensor theory and expressed as,

$$S_p = M * g + \exp(-g) \tag{8}$$

5.2   Improved Exemplar Patch Selection

It is the identical patches that are located in the source region that populate the patch that has the greatest priority with the greatest priority value. Utilizing the correlation measure known as the SMD between the patches of the source area and the patch that has the greatest priority is the method that is utilized in order to accomplish the process of picking an exemplar patch from the source region.

$$\Psi_{q'} = \arg \min_{\Psi_q \in \Phi} d_{SMD}\left(\Psi_p, \Psi_{q^i}\right) \tag{9}$$

The variable Ψ_p represents the highest priority patch, while Ψ_{q^i} represents the exemplar patches of the source region. The SMD is determined by subsequent equation:

$$d_{SMD}\left(\Psi_p, \Psi_{q^i}\right) = \left| \frac{\sum \overline{N} \Psi_p}{\sum \overline{N}} - \frac{\sum N \Psi_{q^i}}{\sum N} \right|^2 \tag{10}$$

The binary mask, referred to as N, is used to indicate absent content in a picture. The expression $\frac{\sum \overline{N} \Psi_p}{\sum \overline{N}}$ computes the mean of the pixels that are currently present in the target patch. Expression $\frac{\sum N \Psi_{q^i}}{\sum N}$ computes the mean value of pixels that will be utilized for filling. These pixels are situated at undisclosed points within the exemplar patch.

As stated in reference (3), a small value of SMD indicates a high similarity between the current pixels in the target patch and the pixels that will be utilized for filling. In contrast, a big value of SMD indicates a significant disparity between the present pixels and the pixels that will be utilized for filling. When employing the exemplar patch to repair the target patch in this scenario, the two components of the repaired patch exhibit significant dissimilarities, hence increasing the likelihood of a mismatch error.

6   Experimental Results and Discussion

A system with an Intel Core 3 processor, 12 GB of RAM, and a 2.7 GHz clock speed is used for the experiments, which are conducted using the MATLAB software. By using it on images extracted from the Berkeley Segmentation dataset, we can confirm that the suggested method is efficient [28]. We compare the results with the state-of-the-art methods to conduct a qualitative analysis. Figures 5, 6, 7, 8 and 9 display the comparison results for five distinct images. The input image, object mask, results from Criminisi et al. [9] and Janardhana Rao et al. [16] are all displayed in each figure. (e) Findings from the study by Yao F et al. [15]; (f) Findings from the approach that was suggested. The artifacts related to the object that was removed are marked in the box. The results clearly show that the proposed method outperformed the existing methodologies in terms of quality when compared visually to the previous works.

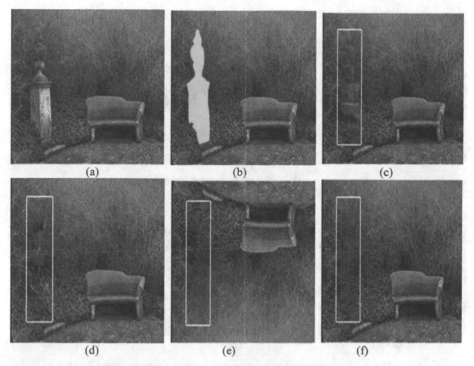

Fig. 5. The outcome images correlation for the sample image-1

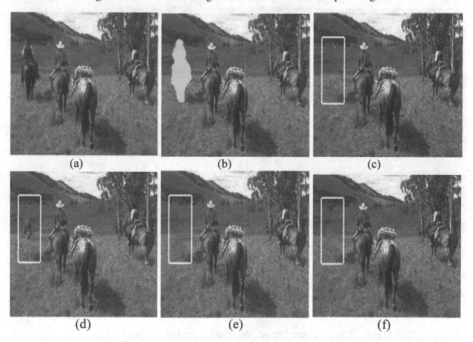

Fig. 6. The outcome images correlation for the sample image-2

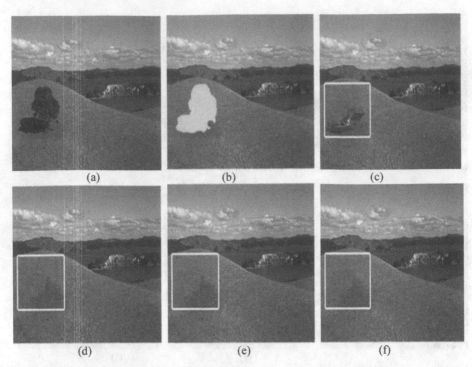

Fig. 7. The outcome images correlation for the sample image-3

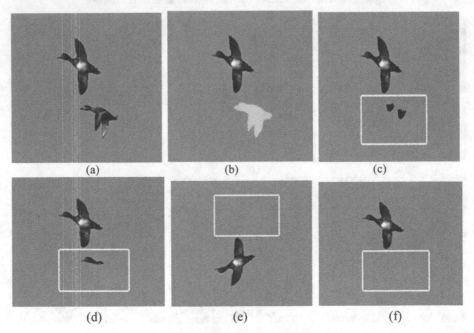

Fig. 8. The outcome images correlation for the sample image-4

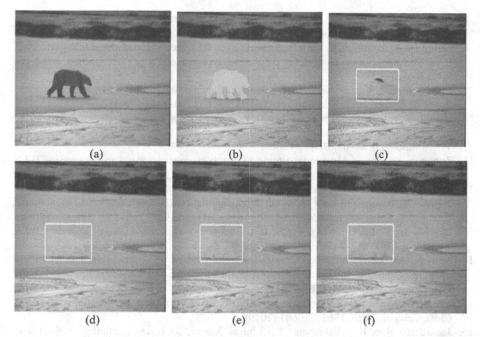

Fig. 9. The outcome images correlation for the sample image-5

7 Conclusions

The upgraded exemplar-type inpainting approach for images is created by introducing the efficient highest priority calculation approach and a novel correlation testing technique. This upgraded technique prevents the dropping effect in the largest priority computation. It uses Square of Mean Difference (SMD) for selecting exemplar patches in the source region. The integration of these both improved techniques yielded highly effective image inpainting outcomes. The achieved inpainting outcomes demonstrate superior efficiency when compared to existing approaches in the literature.

References

1. Rathish Kumar, B.V., Halim, A.: A linear fourth-order PDE-based gray-scale image inpainting model. Comput. Appl. Math. **38**(6), 1–21 (2019)
2. Sridevi, G., Srinivas Kumar, S.: Image inpainting and enhancement using fractional order variational model. Defense Sci. J. **67**(3), 308–315 (2017)
3. Sridevi, G., Srinivas Kumar, S.: P-Laplace variational image inpainting using symmetric Riesz fractional differential filter. Int. J. Electr. Comput. Eng. **7**(2), 850–857 (2017)
4. Sridevi, G., Srinivas Kumar, S.: Image inpainting based on fractional-order nonlinear diffusion for image reconstruction. Circuits Syst. Signal Process. **38**, 3802–3817 (2019)
5. Sridevi, G., Kumar, S.: A qualitative report on diffusion based image inpainting models. Int. J. Comput. Digital Syst. **11**(1), 369–386 (2022)
6. Gamini, S., Gudla, V.V., Bindu, C.H.: Fractional-order diffusion based image denoising model. Int. J. Electr. Electron. Res. **10**(4), 837–842 (2022)

7. Gamini, S., Kumar, S.S.: Homomorphic filtering for the image enhancement based on fractional-order derivative and genetic algorithm. Comput. Electr. Eng. **106**, 108566 (2023)
8. Criminisi, A., Patrik, P., Kentaro, T.: Object removal by exemplar-based inpainting. In: 2003 IEEE Computer Society Conference on Computer Vision and Pattern Recognition, vol. 2, pp. II-II (2003)
9. Criminisi, A., Patrik, P., Kentaro, T.: Region filling and object removal by exemplar-based image inpainting. IEEE Trans. Image Process. **13**(9), 1200–1212 (2004)
10. Huang, C., Chun, H., Sheng, L., Ling, W.: Robust algorithm for exemplar-based image inpainting. In: Proceedings of International Conference on Computer Graphics, Imaging and Visualization, pp. 64–69, Beijing (2005)
11. Zongben, X., Jian, S.: Image inpainting by patch propagation using patch sparsity. IEEE Trans. Image Process. **19**(5), 1153–1165 (2010)
12. Chinmayee, R., Anupama, A., Bagashree, P.: Image inpainting using exemplar based technique with improvised data term. In: 2018 International Conference on Computational Techniques, Electronics and Mechanical Systems (CTEMS), pp. 162–166, Belgaum (2018)
13. Liu, H., Bi, X., Lu, G., Wang, W.: Screen window propagating for image inpainting. IEEE Access **6**, 61761–61772 (2019)
14. Nan, A., Xi, X.: An improved Criminisi algorithm based on a new priority function and updating confidence. In: 2014 7th International Conference on Biomedical Engineering and Informatics, pp. 885–889. IEEE (2014)
15. Yao, F.: Damaged region filling by improved criminisi image inpainting algorithm for thangka. Clust. Comput. **22**(6), 13683–13691 (2019)
16. Janardhana Rao, B., Chakrapani, Y., Srinivas Kumar, S.: Image inpainting method with improved patch priority and patch selection. IETE J. Educ. **59**(1), 26–34 (2018)
17. Revathi, K., Janardhana Rao, B.: Analysis and implementation of enhanced image inpainting method using adjustable patch sizes. Int. J. **9**(3) (2021)
18. Rao, B.J., Krishna, O.V.: Evaluation of image inpainting algorithms. CVR J. Sci. Technol. **7**, 48–52 (2014)
19. Zhang, L., Chang, M.: An image inpainting method for object removal based on difference degree constraint. Multimed. Tools Appl. **80**, 4607–4626 (2021)
20. Abdulla, A.A., Ahmed, M.W.: An improved image quality algorithm for exemplar-based image inpainting. Multimed. Tools Appl. **80**(9), 13143–13156 (2021)
21. Zahra, N., Ghazale, G., Nader, K., Shadrokh, S.: Image inpainting by adaptive fusion of variable spline interpolations. In: 25th International Computer Conference, Computer Society (CSICC), pp. 1–5, IEEE (2020)
22. Ahmed, M.W., Abdulla, A.A.: Quality improvement for exemplar-based image inpainting using a modified searching mechanism. UHD J. Sci. Technol. **4**, 1–8 (2020)
23. Janardhana Rao, B., Chakrapani, Y., Srinivas Kumar, S.: MABC-EPF: video in-painting technique with enhanced priority function and optimal patch search algorithm. Concurr. Comput. Pract. Exp. **34**(11), e6840 (2022)
24. Rao, B.J., Chakrapani, Y., Kumar, S.S.: An enhanced video inpainting technique with grey wolf optimization for object removal application. J. Mob. Multimed. **18**(3), 561–582 (2022)
25. Janardhana Rao, B., Chakrapani, Y., Srinivas Kumar, S.: Video inpainting using advanced homography-based registration method. J. Math. Imaging Vis. **64**(9), 1029–1039 (2022)
26. Janardhana Rao, B., Chakrapani, Y., Srinivas Kumar, S.: Hybridized cuckoo search with multi-verse optimization-based patch matching and deep learning concept for enhancing video inpainting. Comput. J. **65**(9), 2315–2338 (2022)
27. Rao, B.J., Revathi, K., Babu, G.H.: Video inpainting using self-adaptive GMM with improved inpainting technique. CVR J. Sci. Technol. **22**(1), 42–46 (2022)
28. Arbelaez, P., Maire, M., Fowlkes, C., Malik, J.: Contour detection and hierarchical image segmentation. IEEE Trans. Pattern Anal. Mach. Intell.Intell. **33**(5), 898–916 (2011)

Reversible Logic Toffoli Gate Priority Encoder for Effective Nano-Scale Application in QCA Paradigm

K. Kalpana[1]($\boxtimes$) (iD), B. Paulchamy[1] (iD), V. V. Teresa[2] (iD), K. Sivakami[3] (iD),
S. M. Deepa[3] (iD), and N. Revathi[3] (iD)

[1] Department of ECE, Hindusthan Institute of Technology, Coimbatore, India
Kalps.tamil@gmail.com
[2] Department of ECE, Sri Eshwar College of Engineering, Coimbatore, India
[3] Department of ECE, Nehru Institute of Engineering and Technology, Coimbatore, India

Abstract. Main objective of line of reversible priority encoder depends on QCA. Among important stages and processes. A cutting-edge kind of nanotechnology is known as quantum-dot cellular automata (QCA). It could be used as the foundation for reversible and digital circuit construction. In this study, a proposal for a simple, reversible, and encoders with the ratio of four inputs and two outputs are considered. It is possible to construct a reversible encoder circuit by making use of the Toffoli gate design, which is simple and very inexpensive. The simulation tool QCA Designer is used in order to evaluate the suggested designs for their level of structural soundness.

Keywords: Gate EXOR · Gate EXOR · Emitting power · Two-way circuit · Priority Encoder

1 Introduction

The performance of devices built restricted since individual element sizes are so small, and these devices also use a large amount of power. Based on many researchers what they are thinking about how to construct complicated logic circuits at the nanoscale utilizing technologies that use very little power. A lower design with higher switching speed, it is required to consider alternatives to CMOS. The quantum-dot cellular automaton (QCA) is introduced as an innovative framework for fabricating a nanoscale apparatus that exhibits exceptional density and operates at terahertz speeds. This paradigm does not need the use of transistors. There are a number of sources [1–3] that provide an in-depth discussion of the experimental characteristics and physical implementations (including metal-island, semiconductor, magnetic, and molecular QCA). [4] Just recently, the world's first functioning on a fundamental product was successfully created. The CMOS technology has a propensity to waste a great deal of power, which is one of its drawbacks. It is possible to avoid the waste of energy that occurs during calculations by making use of the reversible computation, which has been described in [5]. Research back up assertion. The reversible gate is an essential component logic. The scientific community has

© The Author(s), under exclusive license to Springer Nature Switzerland AG 2024
S. L. Gundehommu et al. (Eds.): REGS 2023, CCIS 2081, pp. 205–216, 2024.
https://doi.org/10.1007/978-3-031-58607-1_15

come up with a few different ideas for different kinds of reversible gates [5]. Due to the vast range of functions that it is capable of [6–9], the Toffoli gate sees a lot of action.

Many researchers are focusing their attention on reversible logic circuits at the moment because of its potential applications in nanotechnology, quantum computing, and optical computing. It is unusual for quantum computers to make use of logic circuits that can be erased. The conventional ways of mixing logic cannot be readily applied to the models of reversible circuits. A reversible logic circuit can never work correctly since it always generates invalid results. Products that were tossed through the gate are most often classified as rubbish. It is necessary to pay the quantum cost associated with each. Popular reversible logic gates that in use today are detailed, along with their QCA designs, in this study. This article will provide that is reversible using a logical switch. This technique should not be too complicated. This design approach is implemented to produce an encoder with n bits of data storage capacity. The use of methods like as equivalence, analysis, and modeling is also investigated.

2 Related Works

The use of QCA to generate reversible circuits is becoming more common in published research. The pervasive use of universal switch, one of the most urgent challenges this area of research is the development of an XOR gate that is both quick and effective. In many contemporary designs, normal majority-voting gates and inverters is considered to be typical. It is demonstrated in reference [10] how an alternate method may be used to create XOR gates by making use of a 5-input majority gate. This method has numerous benefits over the traditional method. On the other hand, not too long ago, a compact design of the XOR gate that makes use of the cell interaction approach [11, 12] was presented, and this has great advantages when it comes to obtaining advanced circuit design while keeping complexity to a minimum.

Numerous studies [11, 12] have shown that reversible logic is a paradigm that can be used in the real world for quantum computing. The fact that data may be read and written in both ways is the major advantage offered by reversible logic circuits. There are several publications that go into depth into the various different layouts that are available for reversible logic gates. These reversible gates enable the creation of reversible logic circuits, which are important for quantum cellular automata and optical computing. Both the sequential circuit pattern and the reversible circuit layout have been investigated [12–14, 18, 19]. Both of these circuit layouts have applications in combinational circuits for the operations of addition, multiplication, and division. Applications for reversible logic gates are numerous and include, but are not limited to, the communication industry, low power design, and quantum computing. The nanocommunication methods discussed in [14–16] indicate a significant amount of interest in making use of these reversible logic gates. The paper includes a description of the design of the QCA, the building of the reversible gate, and the simulation results for both designs.

The Toffoli gate may be implemented using QCA in a number of different ways. The Toffoli gate idea was put into action in [17] with the use of four large-area majority gates. In addition to the three AND gates that are also included, the architecture makes use of four MGs, one of which serves as an OR gate. Recent developments have resulted in the creation of an innovative QCA implementation [20, 21] of a Toffoli gate-based reversible priority encoder. This design incorporates a total of five MGs, four of which serve the job of AND gates, and one of which takes on the function of an OR gate. Because there are 64 quantum cells in the design, there is a delay of 1.5 cycles because of this.

This piece of paper is embellished in seven distinct ways. The qualitative content analysis (QCA) is introduced in Parts 1 and 2, respectively. The methodology behind how the QCA technology works is broken out in detail in Sect. 3. Discussed about different logical switches respectively. In Sect. 6, of the simulation studied by comparing and contrasting the various aspects of the different designs that were offered. The suggested circuits are reviewed in Sect. 7, which also includes some concluding views on where the discussion should go from here.

3 QCA Terminology

Fundamental component that has been modified to include two quantum dots and two extra electrons. As can be seen in Fig. 1, a quantum dot is positioned at each of the four corners of the square QCA cell. The QCA method is dependent on the contact that occurs between adjacent cells as a result of Coulomb's force. In Fig. 2, the quantum wire is shown in the form of a grid of quantum cells. The two main gates in QCA are the majority gate and the inverter gate. Figure 3(a) displays the inverter gate, while Fig. 3(b) illustrates the majority gate. These gates are used in the design of the vast majority of QCA circuits. The mathematical expression is Maj(A,B,C), which is written as AB + BC + AC. In the construction of two-input logic "OR" and "AND" gates, one input of the majority gate is polarized to either (P = +1) or (P = −1), depending on the kind of gate [14–16].

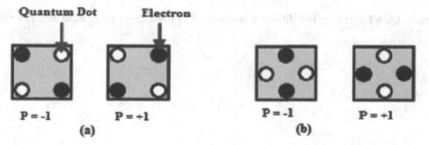

Fig. 1. QCA Cell (a) degree with 90^0 (b) degree with 45^0

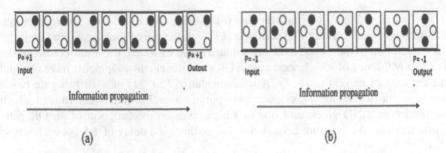

Fig. 2. Connection QCA (a) Using 90 degree cells (b) Using 45degree cells

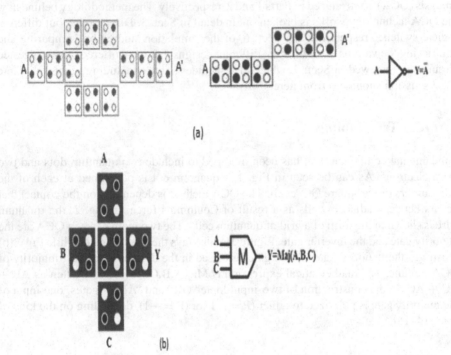

Fig. 3. QCA Layout (a) Two diffrerent layout in QCA Inverter (b) Three input Majority gate

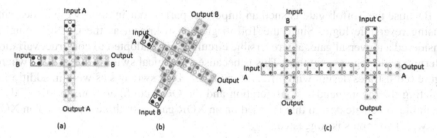

Fig. 4. Crossover (a) Coplanar (b) Multilayer (c) Logical

Figure 4 illustrates the many different crossover techniques that may be used in QCA. Synchronicity in four steps provides synchronicity. Mainly four categories, as shown in Fig. 5, and the "Switch," "Hold," "Release," and "Relax" phases are found each of these clock zones. The propagation of a signal over a QCA wire, which is composed of a succession of QCA cells, is analogous to the transmission of a signal through a conventional shift register [17]. Figure 5 depicts many distinct stages involved the timing process.

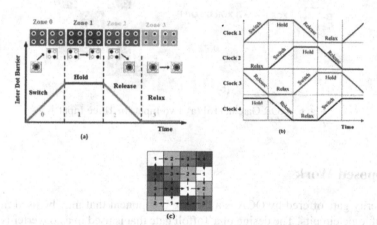

Fig. 5. Timing (a) Zones of timing (b) Phase (c) USE Scheme (Universal, Scalable and Efficient) in QCA

4 Reversible Logic

Because the logic that is being employed is reversible, the system may readily do computations in both the forward and backward directions. If a calculation, and then restarted outcome, it is said to be reversible. The following prerequisites must be met in order for any type of reversible logic to be defined: For all of the data to be kept, there needs to be an equal number of inputs and outputs, as well as a one-to-one relationship between them.

Because the Toffoli gate is such an important part of our inquiry, we will describe it using reversible logic. Since the Toffoli gate, also known as the CCNOT gate, is considered a universal gate, any reversible circuit may be adopted to construct variation in terms of old logical switches. This is because the logical switch is used everywhere. Figure 6 illustrates design with different inputs of XOR switches as well, in addition to depicting the recommended configuration and the QCA configuration. In this study, a reversible logic gate design that is based on an XOR gate with three inputs and an XOR gate with two inputs is suggested.

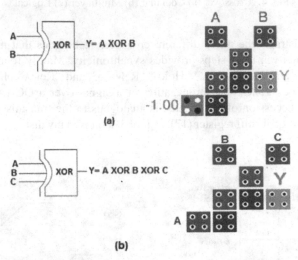

Fig. 6. XOR Gate in [16] (a) Two Input (b) Three Input

5 Proposed Work

The majority gate offered by QCA is a flexible component that may be used in a wide variety of logic circuits. The design of a Toffoli gate that is used in a converter is broken down in depth in this article. Nevertheless, with summation of necessary gate consists of a modest QCA XOR gate as well.

5.1 Toffoli Gate

For the construction of the low hardware overhead QCA priority encoder [12], XOR gates that are based on the cell-interaction theory are used. A Toffoli gate may exist in one of five possible quantum states. This logic gate has three inputs and produces three different results from those inputs. The Toffoli gate may be explained as follows: In the event where none of the initial two bits contain a one, the third bit does not undergo any change; in all other cases, it is inverted. The Toffoli gate, which has the logic equation $R = ABC$, $Q = B$, and $P = A$, has inputs A, B, and C that correlate to corresponding outputs P, Q, and R, as shown in Fig. 7.

Figure 7: The Tofoli gate's equivalent outputs, P, Q, and R, are represented by the inputs A, B, and C. Figure 7 shows a gate called the Toffoli. The format is symbolic of. a) The In-Progress Framework for the QCA. Figure 7b depicts the process that must be followed in order to successfully execute the Toffoli gate QCA illustrated in Fig. 7a. Utilize the QCADesigner tool for both testing and practical application.

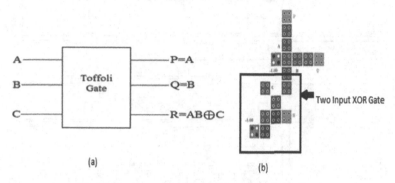

(a)   (b)

Fig. 7. Basic gate: (a) logical form (b) design form

5.2 Gate with Encoder

A system starts with a higher binary inputs and reduces the number of binary outputs. The index of the most significant active line is given by priority encoders in the form of a binary representation, and the counting begins at zero. Due to the high priority with which they respond to interrupt input, they are often used in the management of interrupt requests. To make sure, the necessary component which includes highest preference receives resource when multiple components require it. The converter receives minimum source bits (I3, I2, I1, and I0) and outputs only half of source bits, Y_1 and Y_0. The formula for the expression mentioned below:

$$Y_0 = I_2 + I_3 \tag{1}$$

$$Y_1 = I_2, I_1 + I_3 \tag{2}$$

Outcome of gate simulated was proposed in Fig. 8. To complete construction of a 4-to-2 priority encoder, we have now linked the three Toffoli gates, as seen in Fig. 9. The planned arrangement has six different outputs for garbage. Figure 10 provides an illustration of the proposed architecture for the QCA. The output of circuit, which is two-way always, produce garbage. Nevertheless, seemingly meaningless sinks serve a different purpose in more sophisticated circuits [24–28].

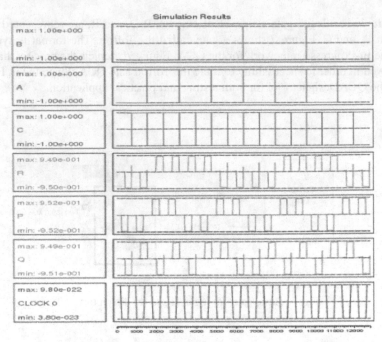

Fig. 8. Results for simulation

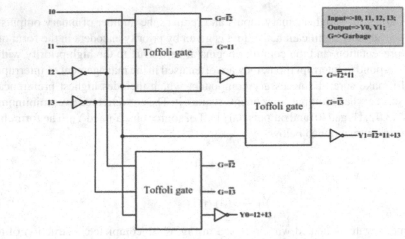

Fig. 9. Reversible 4-to-2 priority encoder block diagram in [21]

6   Comparison and Analysis

In this section, the QCA Designer tool used for an analysis of the circuit's complexity, simulation and evaluation. This proposed work is compared to other work that has been previously published. To calculate QCA logic circuits, this simulation tool makes use of its twin engines respectively. Suggested networks are put through their paces on both simulators in an effort to identify any potential faults. Here, the relevant parameters of both simulation engines are considered. The QCA architecture that was created by the simulator is studied so that the practicability of the resultant values may be determined. The Toffoli gate design that was intended to be used with functioned. Figure 10 and Fig. 11 present results of the simulated, respectively.

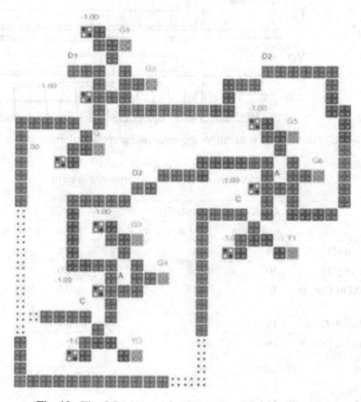

Fig. 10. The QCA Layout for the proposed Priority Encoder

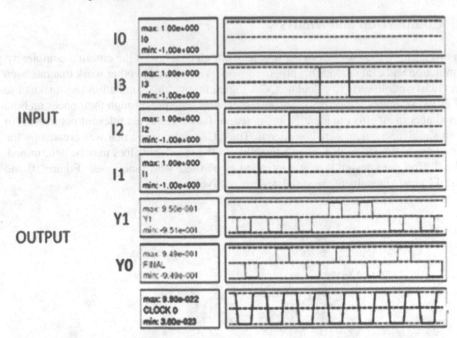

Fig. 11. The Simulation Result for the reversible proposed priority encoder

Table 1. Performance matrices for the proposed designs

Circuits	QCA circuit complexity		
	cell count	Total area (μm^2 Latency)	Latency (clock cycle)
Toffoli gate in [20]	64	0.088	1.5
Toffoli gate in [21]	50	0.058	1.0
Two Input XOR Gate in [16]	8	0.045	0.5
Proposed Toffoli Gate	17	0.048	0.7
Proposed 4-to-2 Priority Encoder	196	0.058	1.0

7 Conclusion

In this work, a 4-to-2 reversible priority encoder is proposed. This encoder is a sophisticated network that consists of gates with minimum number. The gates that were depending on EXOR gate with three inputs includes construction of gate was based on the TIEO. The design that was recommended makes use of fewer cells, takes up has lesser gates than is currently being used. Previous implementations of reversible priority encoders that were proposed for use in QCA have been investigated. The top two articles [20, 21] were chosen to serve as a benchmark for the design that was presented. As can be seen in Table 1, the suggested configuration calls for a lower number of logic gates when compared to the studies [20, 21] that were used as references. As a direct result of this, the amount of time required to do a job is cut down. Recent study [20, 21] found that the recommended designs for the Toffoli gate and the reversible 4-to-2 priority encoder made greater use of the QCA logic gates than the average of the previous research. This improvement was in the range of 87% and 82%, respectively. Table 1 shows that cell count, area utilization, and latency may all be improved with the Toffoli gate QCA configuration. These advantages may be seen in the table. It has been shown that the system benefits significantly by using the suggested method for optimizing QCA circuits.

References

1. Landauer, R.: Irreversibility and heat generation in the computing process. IBM J. Res. Dev. **5**(3), 183–191 (1961)
2. Cho, H., Swartzlander, E.E.: Adder and multiplier design in quantum-dot cellular automata. IEEE Trans. Comput. **58**(6), 721–727 (2009)
3. Zhang, R., Walus, K., Wang, W., Jullien, G.A.: A method of majority logic reduction for quantum cellular automata. IEEE Trans. Nanotechnol. **3**(4), 443–450 (2004)
4. Hänninen, I., Takala, J.: Binary adders on quantum-dot cellular automata. J. Signal Process. Syst. **58**(1), 87–103 (2010)
5. Kummamuru, R.K., Orlov, A.O., Ramasubramaniam, R., Lent, C.S., Bernstein, G.H., Snider, G.L.: Operation of a quantum-dot cellular automata (QCA) shift register and analysis of errors. IEEE Trans. Electron Devices **50**(9), 1906–1913 (2003)
6. Walus, K., Jullien, G.A., Dimitrov, V.S.: Computer arithmetic structures for quantum cellular automata. In: Conference on Record of the Thirty-Seventh Asilomar, Signals, Systems and Computers, vol. 2, pp. 1435–1439 (2003)
7. Walus, K., Dysart, T.J., Jullien, G.A., Budiman, R.A.: QCADesigner: a rapid design and simulation tool for quantum-dot cellular automata. IEEE Trans. Nanotechnol. **3**(1), 26–31 (2004)
8. Lent, C.S., Tougaw, P.D.: Lines of interacting quantum-dot cells: a binary wire. J. Appl. Phys. **74**(10), 6227–6233 (1993)
9. Tougaw, P.D., Lent, C.S.: Logical devices implemented using quantum cellular automata. J. Appl. Phys. **75**(3), 1818–1825 (1994)
10. Ahmad, F., Bhat, G.M., Ahmad, P.Z.: Novel adder circuits based on quantum-dot cellular automata (QCA). Circuits Syst. **5**(6), 142–152 (2014)
11. Garipelly, R., Kiran, P.M., Kumar, A.S.: A review on reversible logic gates and their implementation. Int. J. Emerg. Technol. Adv. Eng. **3**(3), 417–423 (2013)
12. Haghparast, M., Jassbi, S.J., Navi, K., Hashemipour, O.: Design of a novel reversible multiplier circuit using HNG gate in nanotechnology. World Appl. Sci. J. **3**(6), 974–978 (2008)
13. Thapliyal, H., Srinivas, M.B.: Novel reversible multiplier architecture using reversible TSG gate (2006)
14. Bruce, J.W., Thornton, M.A., Shivakumaraiah, L., Kokate, P.S., Li, X.: Efficient adder circuits based on a conservative reversible logic gate. In: Proceedings IEEE Computer Society Annual Symposium on VLSI. New Paradigms for VLSI Systems Design (ISVLSI 2002), pp. 83–88. IEEE (2002)
15. Abdullah-Al-Shafi, M., Shifatul, M., Newaz, A.: A review on reversible logic gates and its QCA implementation. Int. J. Comput. Appl. **128**(2), 27–34 (2015)
16. Das, J.C., Purkayastha, T., De, D.: Reversible nanorouter using QCA for nanocommunication. Nanomater. Energy **5**(1), 28–42 (2016)

17. Sen, B., Dutta, M., Goswami, M., Sikdar, B.K.: Modular design of testable reversible ALU by QCA multiplexer with increase in programmability. Microelectron. J. **45**(11), 1522–1532 (2014). https://doi.org/10.1016/j.mejo.2014.08.012
18. Kasilingam, K., Balaiah, P.: A novel design of nano router with high-speed crossbar scheduler for digital systems in QCA paradigm. Circuit World **48**(4), 464–478 (2022)
19. Kalpana, K., Paulchamy, B., Chinnapparaj, S., Mahendrakan, K., AbdulHayum, A.: A novel design of nano scale TIEO based single layer full adder and full subractor in QCA paradigm. In: 2021 5th International Conference on Intelligent Computing and Control Systems (ICICCS), pp. 575–582. IEEE (2021)
20. Vetteth, A., Walus, K., Dimitrov, V.S., Jullien, G.A.: Quantum-dot cellular automata of flip-flops. ATIPS Laboratory 2500, 1–5 (2003). ATIPS Laboratory 2500 University Drive, N.W., Calgary, Alberta, Canada T2N 1N4
21. Yang, X., Cai, L., Zhao, X.: Low power dual-edge triggered flip-flop structure in quantum dot cellular automata. Electron. Lett. **46**(12), 825–826 (2010)
22. Das, J.C., De, D.: Novel design of reversible priority encoder in quantum dot cellular automata based on Toffoli gate and Feynman gate. J. Supercomput. **75**(10), 6882–6903 (2019)
23. Safoev, N., Abdukhalil, G., Abdisalomovich, K.A.: QCA based priority encoder using Toffoli gate. In: 2020 IEEE 14th International Conference on Application of Information and Communication Technologies (AICT), Tashkent, Uzbekistan, pp. 1–4 (2020)
24. Singh, K., Srinivas, M.B.: Design of efficient reversible logic based priority encoders. In: 2018 IEEE International Conference on Advanced Networks and Telecommunications Systems (ANTS), Indore, India, pp. 1–5 (2018). https://doi.org/10.1109/ANTS.2018.8710105
25. Das, Sarkar, S.K.: Design of efficient reversible logic based Toffoli gate priority encoders. In: 2020 4th International Conference on Intelligent Computing and Control Systems (ICICCS), Madurai, India, pp. 724–729 (2020). https://doi.org/10.1109/ICICCS48265.2020.9124176
26. Sridhar, R., Prathibha, M.: Design and analysis of Toffoli gate based priority encoder using reversible logic. In: 2019 International Conference on Computational Intelligence in Data Science (ICCIDS), Chennai, India, pp. 1–5 (2019). https://doi.org/10.1109/ICCIDS.2019.8724667
27. Singh, P.K., Kadam, R.M., Gaikwad, S.T.: Performance analysis of Toffoli gate based reversible priority encoder using quantum-dot cellular automata (QCA) technology. In: 2017 International Conference on Computational Techniques in Information and Communication Technologies (ICCTICT), Pune, India, pp. 1–5 (2017). https://doi.org/10.1109/ICCTICT.2017.7976777
28. Shridhar, S., Chavan, R.M.: Reversible Toffoli gate priority encoder for quantum-dot cellular automata. In: 2016 IEEE International Conference on Recent Trends in Electronics, Information & Communication Technology (RTEICT), Bangalore, India, pp. 2133–2137 (2016). https://doi.org/10.1109/RTEICT.2016.7808059

Author Index

© The Editor(s) (if applicable) and The Author(s), under exclusive license
to Springer Nature Switzerland AG 2024
S. L. Gundebommu et al. (Eds.): REGS 2023, CCIS 2081, pp. 217–218, 2024.
https://doi.org/10.1007/978-3-031-58607-1

Printed in the United States
by Baker & Taylor Publisher Services

Printed in the United States
by Baker & Taylor Publisher Services